The Archaeus Project

A Historical Novel based on the true story of a group of specialists who secretly work to model a Human's Sixth Sense

T H. Harbinger

Printed in the United States of America

First Printing, 2018

This is a work of historical fiction. Names and characters other than the lead characters either are the products of the author's imagination or are used fictitiously. Any resemblance to actual persons, living or dead, is entirely coincidental. The places, dates, and incidents are authentic and based on records from various archives.

Visit the author's website:
http://www.amazon.in/T-H.-Harbinger/e/B00JEVD256

ISBN-13: 978-1724471529

ISBN-10: 172447152X

The Archaeus Project

Table of Contents

Foreword

Archaeus. The 16th century philosopher and physician Paracelsus first used the term Archaeus to designate "an invisible spirit universal in all things, the healer, the dispenser and composer of all things".

Since that time The Archaeus has been defined as "the hidden virtue of nature", the "invisible sun", and "he who disposes everything according to a definite order, so that each comes to its ultimate matter". In short, "The anatomy of the Archaeus is the anatomy of life".

In December 1982, six professionals came together, drawn by their own experiences with the occult, the paranormal, to secretly research the psychic sciences. They developed their own methods and experiments. The goal was to discover the relationship between the brain, the mind, and the spirit. They called this secret activity the Archaeus Project.

The first six studies hosted by the Archaeus Project were open to the public, meeting on the first Wednesday of the month. Each meeting included experts in the paranormal who shared their knowledge and experiences. A mysterious feeling, and sometimes an odd presence, seemed to spontaneously occur and then dissipate at some point during each meeting. This brought fear to the organizers of the Archaeus Project and further meetings literally went underground. The Archaeus Project moved its meetings into a newly constructed vault in the basement of the West Winds Mansion. It is unknown what these meetings involved and what was discovered. It is unknown if the Archaeus Project continues meeting and experimenting to this day.

This book captures the Project's early findings through archived materials that were made available by the Archaeus Project before they disappeared underground, and through personal experiences. These include a number of videotapes (VHS format) and audiotapes recorded in 1982 and 1983 when the Archaeus Project was being organized. The Archaeus Project also produced a few sporadic publications with this disclaimer:

> *WELCOME to an issue of ARCHAEUS. If anything other than the Archaeus idea unites the content of the articles in this issue, it is that they all discuss what the authors have actively been doing in very recent months: hands-on involvement, if you will. ARCHAEUS is not a journal of mainstream science. For the most part, the facts presented are controversial, and considerable work remains to be done before any of these ideas become firmly accepted and noted in the pages of traditional science—whatever that was or will be.*

In 1983 I had the opportunity to attend some of the Archaeus Project meetings open to the public. My friend and colleague, Professor Otto Schmitt, invited me. For years he had visitors to his laboratory at the University of Minnesota, and he told me some of the strange experiments that these visitors attempted to perform. Not all of them worked as planned. Otto accepted the limited experimental success as a necessary step in studying such a diverse and complex subject as the paranormal. His focus was on extrasensory perception or ESP, and he kept an open mind on its possibilities throughout the remainder of his career. Certainly, there was and still is much controversy about ESP. Because of this, the Archaeus Project did extensive background research to separate psychic hoaxes from authentic ESP events before experts were invited for studies on specific topics. They were well aware that the topic of the paranormal was viewed with much skepticism. Otto had said that the best way to sort out the fake from the true was by witnessing them. Magicians needed thousands of hours of practice to make their magic convincing. People with ESP experiences had little or no practice and described their encounter with a confidence that came naturally.

The study of the paranormal continues to be a lively topic for millions of people worldwide. Nearly everyone has had some type of déjà vu moment. Some people still experience new forces and worlds within our world, only because they have that ability. The Archaeus Project tried to dispel these concepts and focused on the mind itself. The unconscious mind was a difficult thing to test and understand. And parallel to the Heisenberg Uncertainty Principle, the more the Archaeus Project tried to pin down ESP, the less they knew of where it was coming from or going to.

Chapter 1 -- Haunted

West Winds Mansion
Minneapolis, Minnesota
23 April 1929

It has to be safe. It has to be a safe place. It has to be a safe place for children. William Goodfellow was making these points to his architect, Carl Gage. They were standing at a makeshift table of unused lumber in what would become the Great Hall of Goodfellow's five-year dream for this glorious home. It would be a home that would bring him great happiness. The selling of his dry goods business to George Dayton twenty-five years earlier had made him independently wealthy and somewhat happy. He, however, still longed for daily happiness, even a happiness that could be relished minute by minute. William Goodfellow had his eye on a woman he had secretly admired for years and had recently engaged in activity that would lead to making her acquaintance. He needed an opportune moment with her. Maybe this house, this mansion, this home would be the catalyst. He hoped so. He thought of nothing else.

Carl Gage was getting a little tired of all this discussion. He had a mansion to finish. It had grown to fifteen rooms and eleven bathrooms. He had fortified the windows and added extra doorways to every room. Goodfellow stood with Gage in what would become the main banquet room of the mansion. They had fashioned a worktable with oak planks and sawhorses. Gage was annoyed by this discussion and its likelihood of leading to more design changes. However, he knew these changes could give him the opportunity to enhance the Gothic Revival architecture throughout the mansion. Goodfellow was wary of Gage's hidden motives and wanted to have more influence on how these changes took effect.

Goodfellow placed what looked like a newly purchased book on the suspended planks between them. Gage walked around the table to stand to the left of Goodfellow so he could get a clear view of the book. The book, sent to Goodfellow from a friend in New York City, seemed an omen for guidance in building his most prized creation. The book was titled *Rosicrucian Principles for the Home and Business*, written by Dr. H. Spencer Lewis. Goodfellow opened to page 59 and used his finger to underline each word as he read it aloud[1].

> So much for mental alchemy—the art of mentally creating and visualizing the thing desired. Nothing pertaining to the desire must be overlooked. One must keep in mind the usefulness of the thing desired. One must avoid attempting to create that which has no practical use and which will bring neither profit nor benefit to anyone. The dangers from its misuse must be considered and provided against during the process of creating. Its benefits to others must be included, and it must be so created that it cannot fail to be useful to others, and to be of benefit generally to humanity. It must be so created or of such a nature that when it comes into final manifestation objectively, it will become no charge upon the happiness, peace, health, and contentment of others, nor carry with it the sacrifices and sorrow of others. It must be desired with as little selfishness as is consistent with the need of the thing. Its possession must be inspired by no motive associated with revenge or anger, hatred or jealousy, pride or arrogance. Its development and growing reality in the consciousness of its creator should continually inculcate a sense of humility and humbleness, for as it comes into form, the magnificence of the creative powers of man should make the creator of each thing realize his obligations to God and his attunement with God's

[1] Lewis, H. Spencer, Rosicrucian Principles for the Home and Business, Rosicrucian Press, San Jose, CA, 1929.

kingdom. If all of these things are considered, and made a part of the process, then truly one may feel that success and satisfying realization are imminent and assured.

As Goodfellow finished reading the last line, Gage walked to the middle of the room. He turned to stare directly at Goodfellow. He observed Goodfellow's neat appearance, specifically the well-kempt mustache and trimmed, straight hair. This differed considerably from his own appearance, which was scruffier, based on the notion that a half kempt beard and long hair helped bear the winter cold. Gage then began his response. First, he said the whole thing sounded pretty cosmic, calling out the words that seemed to stick in his mind. Alchemy. Desire. Danger. Misuse. Humanity. Happiness. He stopped there, remembering one of the underlying purposes, and made a short comment that this new mansion was to bring not just mental but also physical happiness to William. Goodfellow stood straight up when he heard this, smiled a bit, and encouraged Gage to continue. Sacrifices. Possession. Inculcate. God. Gage turned his back to Goodfellow and walked toward the makings of a giant fireplace on the west wall. He spent some time pointing around the room and talking in phrases about combining certain Pagan and Gothic symbols that should create an aura of openness to new ideas, or at least a curiosity for investigation. Neither man could anticipate how much mystery it might also bring.

(Two years later)

West Winds Mansion
Minneapolis, Minnesota
1 March 1932

She had been given the nickname, Ruby, by her closest friend years ago because she always carried a few of the red stones with her to dispel unhappiness or even pain. She liked the name and now introduced herself this way. Ruby was attracted to the job opening posted by William Goodfellow for a household caretaker. She had grown up in Northern Minnesota on a family farm with pigs, chickens, horses, and cows. They raised everything they needed to eat and sold what they did not need. Her parents were of Scandinavian descent and their three daughters, of whom Ruby was the oldest, were the family's first descendants born in America. During her early years she always felt connected to nature, the farm animals, and her heritage. She belonged and found comfort in this environment, this culture. As she matured, her parents encouraged her and gave her more responsibility. Her neighbors respected her and called on her for help and advice. She built her reputation as a caretaker and had confidence in all she did because she could sense things others could not. She had a way of intuiting things that others seemed not to have. When she saw the job opening to provide a welcoming, secure environment at a new mansion in the southern part of Minneapolis she knew it was the perfect move for her. When Goodfellow met her and contacted her references he knew he had found the best possible choice, and so far, he had no reason to doubt his decision.

Ruby was sitting in her favorite chair in the library. The library was the farthest west room in the West Winds Mansion. She had been hired a year ago by William Goodfellow to be the caretaker and turn his mansion into a home. Goodfellow had gone out for the evening and the cook had just left after cleaning and preparing for the next day meals. It was just after 7pm. Ruby preferred the quietness of the mansion. Its dark wood, many rooms, multiple doors, and its mysterious woodcarvings brought peacefulness to her that few others could understand.

She hears a voice. It's a child. It's a boy. It's a young boy. It's a very young boy. He is shouting for his mother. Mommy. No, it's for his father. Poppy. Did she also hear Charles? Who said Charles? Poppy. Mommy. Where are you? I want my bed. I want my Poppy. Let me go....

Where was this voice coming from? She tried to place it in the room. Was she even awake? She blinked her eyes. Her eyes had been open. She leaned forward in the chair. She was awake and, most likely, had been in her deep meditation state. It was very quiet in this room. It was extremely quiet outside. Usually she could hear the wind whistling through the many leafless trees surrounding the mansion but tonight there was no movement. She considered looking out the window that faced the front garden but the drapes were drawn. There would be nothing to see anyway in the frozen outside world. Cold dark nights were better for creating noise than scenery. She often sat in this room and could envision what was happening on a winter night through sound. This imagined winter scene was more stimulating and exciting than just watching an already frozen landscape freeze more.

She, herself, became frozen in that moment. She closed her eyes and moved her head slowly, like a radar dish, trying to pick up that child's voice again. She began her meditation method. She relaxed her feet and then her legs, freeing them from pressure and weight. Next, she blocked feeling from her womb. Her arms hung limp at her side with her hands purposely resting on her legs, away from her womb. She slowly rotated her head hoping some signal would come into her blank mind. The wind increased producing noises from all directions. She heard a low whistle from the treetops; a low pitch rumbling from bushes scrapping the west wall; packets of plopping snow from swaying branches. Each sound magnified, random, yet united. She had the feeling she was outside in the garden on the north side of the mansion. Each noise made her jump, the plops behind her, the scrapping beside her, the whistling above her. Was she completely surrounded? She heard and processed them all.

She was startled out of her trance-like state by loud noises coming from the front entrance. She knew it was her employer, the mansion owner, William Goodfellow. She could tell by the sounds that the front door was being closed and locked. She knew the sequence of sounds he made as he hung his hat and heavy coat on the coat rack beside the stairs leading to his bedroom on the second floor. His signature, one-step-at-a-time, with a hurried and then slow pace could be heard throughout the house. She wondered if he purposely used this to announce his presence to all living and non-living things throughout the mansion.

She waited for the noises to go all quiet upstairs and then tiptoed to the other end of the long mansion toward her bedroom. It was a small room across a narrow hallway from the kitchen. She hoped she could dream more about the child's voice she'd heard and find an explanation.

West Winds Mansion
Minneapolis, Minnesota
10 April 1932

For the past five weeks William Goodfellow had been obsessed with the kidnapping of the Lindbergh baby. The kidnapping of Charles Lindbergh's 20-month-old son from his own bedroom just after the child had been put to bed had shocked nearly every person in the country. Every newspaper printed a front-page story daily and even President Hoover personally organized the crime investigation at the federal level. There were many leads about the whereabouts of the young Charles, but they all were quickly dispelled. A large ransom, placed into a custom-made box, had been paid using gold certificate bills, with the serial numbers recorded. But surprisingly this produced no truthful leads to the baby's whereabouts. Six months from now a few of these bills would come to Goodfellow's business in Minneapolis.

Goodfellow had asked Ruby to join him after his breakfast in the small dining room off from the kitchen. She had cleared away the breakfast dishes and joined him sitting across from him. Goodfellow began a long dialog on his vision and purpose for the mansion. The mansion name, West Winds, was chosen because winds from the west were the mildest and most favorable. He felt these winds would bring freshness, an unaltered expression from the open prairie. He had sought for the perfect caretaker, and he nodded at Ruby, who would create a sense of comfort and safety in this sturdy mansion. He hoped the right atmosphere and presence would bring him expanded happiness, even a future family. But the outside world was against him.

Twelve years of prohibition had only produced more crime and police corruption. Gangs and mafia organizations lived by their own rules. In escalating their terror on society, they had recently engaged in kidnapping the children of wealthy businessmen collecting thousands of dollars in ransoms. Now with the extensive news coverage of the Lindbergh baby kidnapping everyone was experiencing a kidnapping as if it was his or her own child. How helpless and sorrowful one felt each day. How much strength it took each day to just say that your child was still alive. How disorganized the police forces were. The police even began to use their close connection to the underground crime world for clues. How could a baby be kidnapped from a second story room while the parents and nanny were there? Goodfellow now knew that a mansion was not off-limits to criminals. The meeting ended with Goodfellow asking Ruby to report any unusual activities around the mansion, and suggest ideas of how to increase the security.

That afternoon Ruby was wandering the mansion reflecting on the fear and sadness from this morning's meeting. As she thought more and more about the mansion and having a child kidnapped, she was drawn back to the small dining room. Overwhelmed and numb by sadness she felt detached from the room, from the mansion, from the world. But ever so faintly she began to smell something. What was that smell? It was not the smell of food. It was more like the smell of a forest. Not a fresh, growing forest, but an old, decaying one. It was unpleasant. Ruby became more aware of her surroundings. What was that smell? Where was it coming from?

Ruby took a deep breath and composed herself. She remembered the terrible smell and began to search for it. It would be dreadful if Mr. Goodfellow were to come across it. She called the cook to come into the room. She asked him if he smelled anything unusual. He said no. She asked specifically if he could smell something bad, like a dead mouse or rotting flowers. He said no, there was no such physical smell in that room.

West Winds Mansion
Minneapolis, Minnesota
12 May 1932

Ruby was sitting in the small meeting room to the west of the Great Hall. This was the quietest room in the mansion. Like nearly every room and hallway, there were bookcases holding a variety of books and magazines on history, medicine, and politics. In this room there were also books and magazines on the paranormal, the occult. She had been attracted to this type of information since her youth when her parents had purchased an Ouija 'talking' board; she loved playing with it from the day it arrived. She would gather her sisters, her parents, friends, and neighbors for a session to reveal unknown answers to questions. She led the discussion as to what topic or question they wanted some advice on. She always got goose bumps when the small pointing platform would begin to move. She was certain no one was controlling it when words were spelled out in response to questions. Ruby did more listening than talking and absorbed the entire activity in a way she could not readily describe.

Tonight, she chose this room for her nightly reading and was paging through a copy of The Occult Digest. There was an article about meditation and some guidelines on how to achieve the most relaxed state. She was following these steps and at one point extinguished all the lights in the room. She sat in total darkness. She had put herself into an almost unconscious state. It seemed a level of awareness that was not coming through her senses. Or was it? She analyzed her senses one by one. There was nothing to smell, nothing to hear, and nothing to see as the room was completely black. She felt nothing in this relaxed state and certainly had nothing to taste. Suddenly her legs began to feel cold. But there was no draft. The room was sealed and the temperature outside was nearly the same as inside. She experienced this cold sensation for some time, shaking her legs trying to get her blood moving. But the cold feeling persisted. Slowly she started to feel electricity flowing into her arms, even though her blouse covered her arms. This electricity had a strange feeling. It reminded her of the goose bumps she got when the Ouija platform started moving on its own. Her arms felt like there was static electricity surrounding them.

Ruby continued to welcome these sensations and maintained her relaxed state. Then ever so slowly a white outline appeared to her. It was coming from the far corner of the room. She wanted to see it more clearly. She closed her eyes, then opened them again. This faint object remained and seemed to be approaching her. -It was taking more of a shape. It was a young boy in a one-piece sleeping suit. He was reaching out for someone to pick him up. He slowly spun to the right and revealed a crushed area on the side of his head. Ruby opened her eyes and closed them several times. The image was always there. She wanted to help this boy. She grabbed for the light and snapped it on. The bright light overwhelmed her eyes. There was nothing to see now. The vision and feeling had disappeared. Ruby could not explain the presence she felt but knew it was real. It remained clearly etched in her mind.

West Winds Mansion
Minneapolis, Minnesota
13 May 1932

There was a rumor yesterday, and today the newspapers confirmed it. The Lindbergh baby was dead. He had been found in a wooded area only a few miles from his home. He had probably been killed within a few hours of being kidnapped. The entire nation was grief-stricken. It was as if everyone had lost his or her only child. Charles and Anne Lindbergh were so devastated they wouldn't talk with anyone. It was a day of hopelessness. It was a day that put a stamp on the lawlessness in America.

Ruby found Goodfellow sitting on one of the two William and Mary styled chairs in the sitting area to the west of the front entrance. His body was limp with his chin on his chest and tears were running down his cheeks. Ruby sat in the other chair placing her right hand on the small table between the two chairs. She joined him in the misery holding back an outpouring of sobbing. Eventually Goodfellow raised his head and slowly placed his left hand on top of hers. He did not speak. Ruby collected herself and started to speak softly. She reminded him that he had requested her to mention anything unusual. Then she told him all the things she had heard, felt, seen, and even smelled; things that seemed related to the Lindbergh baby kidnapping. She said she was not sure how real they were but they were very real in her memory. She described the night in the library when she heard a boy's voice with no other noise present. How a couple weeks later she smelled a rotting odor in the dining room when there was no explanation for such a smell. And then just last night, in the small meeting room, she had felt and had seen an apparition of a small boy reaching out for someone to hold him.

Goodfellow looked up and inhaled deeply, almost as if he was gathering new strength. He asked her about each sensory experience and questioned many aspects that could have caused Ruby to falter. Instead, she related each experience or memory in such detail that after thirty minutes of inquiry Goodfellow accepted that what Ruby experienced was very real to her. It was during this questioning that Goodfellow began to formulate an idea, a clear purpose for the future of his mansion. He saw that the experiences of a World War, the consequences of Prohibition, the results of a stock market crash, and the presence of the uncaring President Hoover exposed the total lack of concern for children. He rationalized that Ruby's extrasensory experiences were evidence of her gift for being a caretaker. This mansion had a tremendous potential to become that safe haven for children of all backgrounds and situations.

West Winds Mansion
Minneapolis, Minnesota
1932-1942

With this renewed vision, William Goodfellow increased the security of West Winds Mansion. A high wall surrounding the property was built. Many types of 'booby-traps' were added to the house, not so much to injure but to trigger an alarm. Most would create a startling sound as lightly balanced objects fell to the floor. Goodfellow never mentioned if these booby-traps were intended to surprise intruders or reveal the presence of spiritual visitors. Ruby continued to experience extrasensory perceptions in many rooms of the mansion. She had numerous encounters and was able to connect many of her experiences with abused children she read about in the news.

Goodfellow became so focused on creating a safe haven for children, he lost sight of his original purpose for the mansion, to attract the love of his life and begin a family. He had maintained a passive approach toward this self-absorbing desire. His intentions were to avoid raising even the slightest chance for any fear or uneasiness. Unfortunately, under this non-aggressive environment his love slipped away. When he came to realize this, his view of the future changed. He decided the mansion should be dedicated as an open space for children in anguish. Ruby would continue to care for and take in children and families temporarily in their time of need. Goodfellow realized that his presence could be seen as a deterrent to those needing to stay. By the mid-1930's Goodfellow had moved to a smaller living place and encouraged the full use of the mansion under Ruby's caretaker guidance.

The West Wind Mansion provided a safe place for the community. It was never advertised as such. Because of its nature to protect children it kept a low profile in the news. In 1944 William Goodfellow passed away. Ruby honored his final request that the mansion continue to benefit children. Shortly after his death, it was donated to a Girl Scout organization that would benefit thousands of children in the community for decades to come.

Chapter 2 -- The 'Ghost'

(Fifty years later)

West Winds Mansion
Minneapolis, Minnesota
1 December 1982

Earl Bakken had invited five special guests to this old mansion. As they arrived, each found his or her way into the study located in the southwest corner on the main floor. They did not know each other that well. They all came from different parts of the Twin Cities and had widely different occupations. Little was said as each new person entered and took an empty chair around the large oval shaped dark oak table. The room was overly warm although the outside temperature was well below freezing and a cold northwest wind was slapping the windows and shutters. Noticing the warm temperature in the room, each would leave the door open hoping to let some of the warm air escape. But just as the new person sat down at the table, the open door would gently close as if the cold air of the hall did not want to interact with the warm air in this study. Everyone noticed but was not alarmed. The five guests filled the open chairs at the table with two facing the inside east wall and three facing the outside west wall. The chair at the head of the table farthest from the hallway door was kept empty.

The room had a strange style to it, as if it tried to be different from the rest of the rooms in the mansion, or any other room in the world for that matter. On the west wall were five floor-to-ceiling windows that opened to the possibilities of the west, the frontier. There were doors on every wall with two doors on the east wall, both opening to separate sides of the Great Hall. A fireplace was angled in the northeast corner and seemed to have a self-regulating fire that persisted all night. There were a few strange things hung on the walls and set on small tables around the perimeter of the room. The ceiling was over ten feet high and displayed an unusual pattern that could not be matched to any known pattern. All the guests felt comfortable in the silence. Although each appeared relaxed, their minds were filled with hundreds of thoughts. Wasn't this mansion haunted? What were the stories about this place? Who were these other people? Everything should be making me uncomfortable yet where was this peaceful attitude coming from? Should I say something? What would I say?

Their attention was suddenly drawn to the hallway door; all eyes fixed on it the second before it opened. The door seemingly opened by a puff of wind and in walked their host, Earl Bakken. Everyone stood simultaneously to greet him, but Earl walked past them to the open chair at the head of the table. They sat as he lowered himself into his chair. He looked at them one by one. Earl didn't need an introduction. He was well known around the city as inventor, entrepreneur, and business executive having commercialized the first implantable pacemaker. His 15-year-old company, Medtronic, had now sold over one million devices. He then began talking to the group, but only directed his attention to the open end of the table far from himself. He first thanked them for prompt attendance at the meeting in what was now called the Bakken Library on this Wednesday night. He reminded them that Wednesday was the day of the middle, a day for linking old and new, a day to communicate and plan. Ancient philosophers designated this as a day meant to step into one's true potential, intuition, and emotions. From today on they were bound as a group, a group that had great potential, a group that would learn from each member's intuition, and a group that would test everyone's and anyone's emotions. Each of the five guests had a slight reaction when the word 'anyone' was pronounced. Each had a different thought race through his or her mind. Their attention came back to the table as Earl finished his introductory remarks by reminding them of the founding mission for this group:

To study current and historical processes and techniques for the detection and measurement of bioenergetic fields

*and to apply the results to the realization of human
potential and the diagnosis, treatment, and prevention of
disease.*

He reminded them that he had been working with many
individuals for the past six months. During that time, he had
formulated this mission and at the same time selected a chosen
few to participate in this new quest. This compiled mission
statement was noble but it lacked specifics for taking action. Earl
now wanted a focused topic, a topic toward which they could
direct their winter studies. He had already mentioned the topic
previously to a few members but wanted to have it accepted as
an official focus. He didn't know how to introduce it except to
just say it. Their efforts and meetings should focus on the study
of Extrasensory Perception, ESP. Earl then wanted each member
to share his or her thoughts on the topic of ESP.

To his right was Dennis Shilling. Dennis was considered by some
to be a genius, although he had limited formal education. He was
an avid reader of the literature of the paranormal and the occult.
His favorite book was <u>The Encyclopedia of Ancient and
Forbidden Knowledge</u> by Zolar, in which he had noted
something on nearly every page related to his own experiences.
Presently Dennis was a janitor at the university. Earl had met
him some years ago in a hallway of the medical building and was
immediately struck by Dennis's insights into the mind and body
interactions. Since the purchase of this West Winds Mansion,
Earl had hired Dennis part time to organize and catalog his
extensive medical device collection. Dennis admitted it took a
mansion to hold and display the vast collection. And he felt the
mansion was an appropriate place to house these exotic items.
He pointed out one device that was hanging above the fireplace.
It was called the Heidelberg Electric Belt. It sold for the
outrageous sum of $18 in the 1920's and plugged into the
Alternating Current being brought to homes by the utility power
companies. The device came with several features, the most
popular was advertised to bring youthful vigor to a man's
manhood. One inserted his member through a tight-fitting

leather band that was filled with coils of wire. The alternating current penetrated and permeated the affected part, vitalizing nerves and strengthening tissue. Simply put, it was an electrotherapeutical impotence cure, "shocking the penis 'til it worked good again". All the men in the room slipped their hand under the table as Dennis spoke of this. He described this device with such clarity and devotion that it made these visitors think Dennis had a mental disorder, but they realized he just spoke with extreme openness and honesty, and maybe from his own experience. Earl stated that Dennis would be the secretary and document all the activities related to this group's meetings. Dennis wrote this into his three-ring binder.

John Hayward, to the right of Dennis, was introduced next. John was a priest, which was also evident by his white-collar attire. He had kept both his hands visible on the table during Dennis's introduction. He decided to become a priest in the 1970's after growing up near Necedah, Wisconsin, a town made famous by visitations of the Blessed Virgin Mary. As a teenager, he vividly remembers Mrs. Van Hoof's eight visits with the Holy Mary that started on November 12, 1949. The peak of the excitement came on August 15, 1950 when the Blessed Virgin Mary spoke through Mrs. Van Hoof to a crowd of 100,000. He was fascinated with the precision with which the Holy Mary predicted each return, and then how Mrs. Van Hoof quietly shared those dates that led to thousands of people flooding the farm. Only Mrs. Van Hoof could see the Blessed Mary. But immediately after these private meetings, Mrs. Van Hoof repeated in great detail what was said often using a stage and microphone. John's memory was fuzzy on what was said by the Virgin Mary, but it seemed to revolve around the concept that people should pray more. As he was studying to be a priest he learned that the Catholic Church did not recognize these Necedah visits as miracles although they fit the format of a Godly apparition. The church chose not to include these events in their recognized list of saintly visitations. However, Mrs. Van Hoof maintained her miraculous experiences up to her death, which was earlier this year. Many others who believed have established a complex just east of Necedah where the visitations took place. Shrines have been created to honor each of the eight visitations of the Blessed Virgin Mary.

John started to ask some questions. Was Mrs. Van Hoof just seeing ghosts? What are ghosts? Are ghosts different than visitations from God? Why do people have visitations from the afterlife? Why is the afterlife so important to religion? John went on to say that he didn't have answers to all these questions. Of course, his training as a priest had provided him dogma on heaven and hell and other stories of the afterlife. The Church's teachings were clear on what constituted a recognized visitation from the Virgin Mary, but he thought there was a more complete method that could include a scientific approach. He hoped that this group could help him gain a new understanding to this complex situation.

Sitting across from John, at the far left end of the table from Earl, was Wendell Patel. Wendell was a local banker having moved from the East Coast to expand the family business. He was a third generation American having family roots in Punjab, India. He practiced yoga but was more interested in understanding what he called 'the noises of the beyond'. He defined yoga as removal of the mind's noise so that one could return to a view of the Self. But his experiences seemed to take him beyond the Self into the undefined. As a banker, he saw that the need for security led to the confinement of information, which changed the understanding of the way things were. Wendell had learned that truth was a higher calling than security, and the pursuit of truth required one to challenge his or her accepted worldview.

Wendell was especially interested in spoon bending. He had witnessed many people, from all facets of life, practicing the art of bending metal objects. The spoon bending participants had become convinced that something extraordinary was happening. He felt that this phenomenon had profound and tangible implications for modeling the relationship of mind to world. He expressed that there was no question that "hard" science has progressed to the point where it could describe everyday reality very well, but what about human consciousness? For the past seven years, Wendell had focused his free time on understanding and explaining paranormal phenomena related to consciousness, especially psychokinesis. He then offered several phrases:

Imagination takes over when information stops.

It's not WHAT a person believes, but WHY he believes it that marks the truth of his thinking.

Pooling ignorance does not produce wisdom.

Wendell was talking slowly allowing his words to be processed one at a time. After processing the word wisdom for some time, Earl quietly noted that more effort would be needed than just the pooling of the six of them together.

To Wendell's right was Karen Ness. Dr. Karen Ness, she corrected Earl. She was a doctor specializing in Behavioral Pediatrics at the Minneapolis Children's Hospital. She was especially interested in exploring the use of computers and modern electronic sensing to treat various physiological problems without medications. She launched into her introductory remarks with the phrase: Most adults don't connect the act of thinking in a certain manner with a physical response. The possibility of reversing one's thinking habits, and concurrently reversing a negative physiologic process, was not a part of Western medicine. Spontaneous images were similar to dreams but resulted from different aspects of the brain. Dreams were defined as forming during sleep. When you were sleeping your conscious mind slept, but your unconscious mind was awake. Everyone has had dreams. But it was rare that people had spontaneous images. Spontaneous images seemed then to be connected to the conscious as well as the unconscious mind. She wanted to better define the person who seemed more capable of spontaneous images. In her experience it seemed these people were drawn to the profession of a medium or a healer. There was much yet to understand about healers and the senses they possessed that seemed far above the normal. What really was imagination? This was an important question. Was imagination the only source for creativity? Wasn't creativity the fundamental source for innovation? And wasn't innovation the best source for success? Before Earl could gently end her introduction, she turned to Earl and asked if Earl might not be a good study for this himself. Sessions should be devoted to the evolution of his thinking starting with the great success of his pacemaker company back to the spontaneous image that launched it.

Earl nodded slightly at Karen's suggestion and then introduced the last person. On Earl's left was Otto Schmitt. Otto, and his wife Viola, had come to Minnesota in 1947 to establish the first Biophysics Department at the University of Minnesota. For over 35 years they had devised, developed, and performed experiments to measure electromagnetic effects on the human body. In his publications, Otto had reported data showing a normal statistical distribution for the human characteristic being studied. But in nearly every study they had recorded outliers, subjects whose responses were far outside the norm, almost to the point that it took something other than scientific imagination to accept it.

Otto had taken as a premise Descartes' description of the human body where the body worked like a machine, that it had material properties. Since Descartes lived in the 17th Century, an entirely religious age, it was extremely revolutionary to define the materiality of the body as the object of study, and not the spirituality of the body. This thinking would eventually form the foundation to the philosophical theme of rationalism. As part of this, Descartes described the mind as nonmaterial, therefore not following the laws of nature, so it was left out of the theory and model he had famously defined. But outliers were still in Otto's data and the Descartes model could not explain them. Statistically Otto could ignore them, but now he felt it time to address them directly. In 1949, the philosopher, Gilbert Ryle, had attacked the Descartes model as a 'big mistake' and referred to the idea of a fundamental distinction between mind and matter as "the ghost in the machine". The mind and body was one entity, not a machine haunted by a ghost. Well this phrase had kept Otto thinking about a model for the brain that included this 'ghost'. He looked at Karen and said he had had a spontaneous image one night. He envisioned a way to introduce this into the model using the concept of ESP, extrasensory perception. He was ready to show the group this model and hoped they would help devise experiments to test the model in pursuit of their own passions concerning the paranormal.

Earl then thanked everyone for his or her introductory remarks. Clearly everyone brought passion and something unique to the table. It was and would be a special group that was about to explore, not the last frontier (playing off his Star Trek devotion) but the closest frontier, the mind. The challenge may be that this frontier was even vaster than space. He was willing to provide the funding for these studies. He only asked for each member of this organization to dedicate his or her time and passion toward the cooperative research. They would share their results with the public only when each member agreed to the findings.

This approach of having a group of experts review and bring a conclusion to the reality and potential of the paranormal had been used before. In 1924 The Scientific American magazine launched a contest offering $5,000 to be awarded for 'conclusive psychic manifestations'. They assembled a small group of prominent judges that included the famous escape artist, Harry Houdini. The contest focused on séances, and the mediums that orchestrated them. They reviewed thousands of willing candidates with many exhibiting amazing demonstrations of conversations with the deceased. After over a year of investigations all were readily debunked except one. The press, which followed this contest with almost daily reports, named the most promising candidate, The Witch of Beacon Hill. The judges attended over forty séances with her. All six judges except Houdini were convinced of her clairvoyant psychic ability. Houdini continued to hold out a confirmation and the award was never given. Earl made it clear he expected this new effort they were about to launch be more like a project than a contest. They should build a set of investigations that would lead to conclusions, and not drive for a specific win-lose outcome.

Earl then described how he envisioned the next few months might proceed. Each month one of them would take the lead to plan an experiment based around his or her main interest in the paranormal. He suspected that this should involve inviting unique individuals to come to the Twin Cities. It might be necessary for some to stay as long as a week. He felt the main activity would occur in the Great Hall in this mansion, which was just through one of those two doors to his right.

Earl then faded into a hushed, less sure monologue. He wondered out loud to himself if these experiments would re-awaken the paranormal activity rumored to have occurred in this mansion. A wealthy businessman, William Goodfellow, had built this mansion over fifty years ago. He had hoped it would bring him the greatest happiness: a loving wife and many children. But that didn't happen. Over the years there had been many rumors of why this didn't happen. Some of the rumors point to Goodfellow himself and his fixation on the Lindbergh baby kidnapping. It consumed his thoughts and interfered with his daily life. He displayed strength but his attempts to fortify the mansion indicated an unconscious state of weakness and uncertainty. Other rumors pointed to the design of the mansion itself and its confusing inclusion of both Pagan and Gothic symbolism in the architecture. There were numerous Pagan-like bird and Gothic-looking rose carvings in the extensive woodwork throughout the mansion. Paganism was linked to the practices of the occult whereas Gothic architecture was the choice of the Catholic Church as it rose in power hundreds of years ago. Was this mansion trying to bring those two worlds back together?

Within a few years of its completion, Goodfellow accepted his fate; this mansion was not going to bring him the happiness he sought. In the end, nothing ever did. Goodfellow never married nor had children. But the mansion seemed to have achieved its own happiness. For forty years caretakers, mostly women, had established and run this mansion as a welcoming and safe place for children.

Earl dropped this historical review abruptly holding back a deep feeling of anguish. He switched to a business-like voice and manner. The challenges of the future couldn't be foreseen, at least not by him. But there were hundreds of documented stories about people having special abilities to feel another person's thoughts and even communicate with the dead. Joseph Banks Rhine had introduced the phrase Extrasensory Perception, nicknamed ESP, exactly fifty years ago with the intent to make the field sound normal and linked to the field of psychology. Perception was an established subject of psychologists. He hoped that the study of Extrasensory Perception could be approached scientifically and not as a nonprofessional pastime. Building on J B Rhine's insights Earl agreed that currently each of them was treating this study of the paranormal as a nonprofessional pastime. The moment had now come for them to combine their professional skills and conduct a planned investigation into ESP. As a successful businessman he knew the best way to organize a new venture was to set a goal, maybe even a lofty goal. With this in mind, he offered his overall goal for this project:

> *Through a better understanding of the mind, or at least a*
> *mind-body model, this project would lead to developing*
> *medical practices that could double the quality of life at*
> *half the cost.*

His final announcement was to give this activity a name. He suggested The Archaeus Project. The 16th century philosopher and physician Paracelsus first used the term Archaeus to designate 'an invisible spirit universal in all things'. Easier put, The Archaeus Project was in pursuit of the 'ghost in the machine', a phrase used by philosophers for the past 45 years.

Chapter 3 -- The Model

West Winds Mansion
Minneapolis, Minnesota
5 January 1983

Otto Schmitt was running late. He had moved to this city 35 years ago and still was not used to the cold and harshness of a Minnesota winter. He was a world-renowned scientist in combining physics and biology. No one had done more experimenting in measuring the electrical and magnetic properties of the human body. His work had provided a scientific basis for the development of hundreds of medical devices, one of which was the implanted pacemaker. This device had revolutionized heart disease treatment and made its inventor a very wealthy person. Earl Bakken had always honored Otto since they had met 35 years ago. Otto and Earl had devised many experiments that resulted in outcomes instrumental in defining the many products of Earl's vast company, Medtronic. Earl certainly owed Otto for the advice and data that had guided so many product developments, and he never forgot that debt. In the same way Otto had always showed gratitude and respect for Earl and enjoyed the challenges that Earl always seemed to be bringing to Otto. This Wednesday night was going to be another chance as these two launched into the unknown one more time.

Otto drove his Ford sedan from his house near the University and traveled south on Calhoun Parkway, turned west on 36th St, and then turned north on Zenith Ave. He had to park north of the West Winds Mansion property because that side of the street already had seven cars parked closer. He tried to make a walk path over the snow bank along the edge of the plowed road. As he stepped his weight caused his foot to crash through the ice crusted snow. He had his rubber overshoes on with his pants tucked inside so he did not worry about the depth of the snow.

He slowly made a series of steps, or holes, into the snow bank as he worked his way over that treacherous man-made barrier. As he stepped out of the last hole he had made in the snow bank he headed cross-country toward the West Winds Mansion. The snow was over one foot deep and, although he was in a hurry, took long deliberate steps to work his way across the yard of snow. It never occurred to him, that at his age, he should have looked for the shoveled path.

He entered the garden area where sculptures were covered in a blanket of snow. He paid no attention to them and headed for the front door. Inside he stopped to listen for any voices in the mansion. He wasn't sure exactly where this meeting was taking place. He didn't hear anything except the outside wind and inside silence. First, he took a few steps to the left and peered toward the kitchen. Then he turned slightly right and peered into what was called the dining room. They were both empty with minimal lighting. He then turned to the west and headed toward the doorway at the other end of the entrance hallway. He took the two steps of stairs up to a small sitting area where there was a large wooden staircase ascending. He continued west and took another two-step stairs through a doorway into a small library. Now he could hear some voices to his left and pushed open the door to the study. The same group of people was sitting around the table exactly as they had been one month ago. Instead of taking his former seat next to Earl he stopped at the end of the table opposite of Earl.

Otto set his briefcase in the open chair at the end of the table and pulled the chair back. He took out a folder of papers and set it on the table as he stood at its end. He said hello to everyone and apologized for being late. Earl made a joke that an emeritus professor should be offered at least thirty minutes of schedule leniency and Otto had cut it close. Earl then reminded everyone that Otto would be leading the meeting tonight, this second meeting of the Archaeus Project. He had asked Otto to set a direction for the studies.

Otto took a deep breath and slowly glanced at each person around the table. Clockwise from his left was John, Dennis, Earl, Karen and Wendell. He then made his opening statement.

"Mass appeal for the paranormal was a result of World War I where over 17 million people, mostly young men, died. The survivors, especially those in Europe and North America, wanted to know that their loved ones had not suffered, and desired a way to communicate with them. In the early 1920's thousands of mediums evolved to hold séances to fill this need to communicate with the dead. The Ouija board became a tremendously popular activity for parties and home gatherings. Many scientific organizations pursued activities to determine if there was a science behind these 'psychic manifestations'. Earl had already mentioned the contest sponsored by the Scientific American magazine offering $5,000 for mediums that could convince a select set of scientists and magicians of his or her psychic powers. In the first few months over 35,000 mediums applied to be evaluated. In the end only one woman withstood the scrutiny, and she also couldn't convince every judge of her spiritualistic powers. These scientific studies of the paranormal continued into the 1930's, but the outbreak of World War II refocused everyone back to the present reality of war. I have to add that the mass appeal for the paranormal continued but no longer drew the interest of scientific scrutiny."

"Is it time to start another scientific investigation into the paranormal methods?" Otto then listed, from memory, the major topics that were associated with the paranormal. He spoke slowly so that his team had time to formulate a picture or recall an experience for each one:

    Ouija
    Tarot
    Astrology
    Colorology
    Dice Divination
    Dreams Interpretation
    Graphology

Numerology
Moleosophy
Palm reading
Phrenology
Physiognomy
Rod Radiesthesia
Pendulum Radiesthesia
Superstitions
Lucky signs
Tasseography

Several wanted to ask questions but they realized Otto's style included a lecture first, followed by discussion. Otto posed a second question, "Should the Archaeus Project investigate experiences associated with the occult or supernatural?" He then listed possible topics in the same pausing style as before, except he looked at one of the project members each time he spoke:

Deep Religious Desires
Ghost Stories
Remote Viewing
Séances
Mind Reading
Precognition
Clairvoyance
Psychokinesis
Telepathy
Conspiracy Theories
Secret Government Organizations
UFOs
Aliens

Otto paused a bit longer as he looked around the room making eye contact with each person. He appeared to be reading their minds for answers to his questions. He then opened his folder and looked at the small stack of papers inside. His years of experience in solving problems revealed two basic categories: puzzles and mysteries. Otto reached into his suit coat pocket and pulled out a Rubik's Cube. He held it out and said that he expected that everyone knew what this was. It was a toy that had been sold now for over five years. It had nine squares on each of its six sides. Each square was colored with one of six colors. Each face of the cube could be turned in 90-degree steps, but only one face at a time. The goal was to mix up the squares and then rotate the faces until each side had only one color of squares on it. It was a great example of a puzzle. A puzzle had a bounded amount of uncertainty and then could be solved by using a process. Already there had been several books written outlining processes to solve this puzzle. Each claimed a new method that could solve the puzzle faster with fewer steps. Otto said that it actually took him nearly an hour to solve it, but some people could solve it in less than twenty seconds.

Otto then stated he felt that a study of the paranormal was more like solving a mystery than a puzzle. He asked the group, "When one thinks of solving a mystery, what comes to mind?" Everyone had an answer, but because they were obeying the style of the academic lecture, no one spoke. Otto then answered his own question by saying Sherlock, Sherlock Holmes. Sherlock used a deductive method to solve the problem, the mystery. Solving a mystery took logic and judgment instead of a data rich process. The deductive method does not come out of thin air. It comes from repeated testing inputs against experience, or even better, a theory. And one of the best ways to define a theory was to create a model that illustrated that theory. Otto took a stack of papers from the folder and passed it to Wendell on his right. He asked them to take one sheet. Once each had the sheet and studied it for a few minutes Otto continued. He reminded everyone that Earl had suggested the project focus on ESP, extrasensory perception. Otto said he preferred to think of ESP as the 'ghost in the machine'.

The term ESP came into general use over fifty years ago. It had become a platform for the collection of thousands of firsthand stories involving supernatural experiences. Recognized all over the world and throughout centuries, ESP had attained some level of a vetting process. It had also lent itself to laboratory testing. The most popular testing was done with a deck of cards designed by psychologist, Karl Zener. The Zener card deck had five symbols: a hollow circle, a cross, three vertical wavy lines, a hollow square, and a hollow five-pointed star. Five cards of each design combined to form a deck of 25. A person conducting the test picked up a card in a shuffled pack and noted the symbol on the card. The subject being tested was then asked to guess which of the five cards was being held. The test continued until all the cards in the pack had been used. Otto described the many problems with this type of testing. His main concern with this approach was that it treated ESP as a puzzle. If the subject got 20 percent right he or she was guessing. If they got more than 20 percent right they were either cheating or had ESP. To isolate the cheating, more tests were run which often led to strange interpretations of how the cards were being presented to the subject. After all these double blind tests there were still cases that could not be explained. In addition, no one quite knew what to make of people who had less than 20 percent. This type of testing assumed that ESP was a continuous, conscious activity. But what if it was more an unconscious or spontaneous activity? This would make ESP more a mystery than a puzzle.

Otto compared the current studies of ESP with early Astronomy. For thousands of years the day and night sky was observed and documented with great precision. Babylonian astronomers kept careful records about celestial happenings including the motions of Mercury, Venus, the Sun, and the Moon on clay tablets dating from 1700 BC. By carefully noting local lunar and solar eclipses, the following generations of astronomers were eventually able to predict lunar eclipses and later solar eclipses with fair accuracy. This centuries-long documentation was also made openly available, including their observation techniques, which then encouraged others to duplicate the results. However, significant progress was not made until models for the sun, planets, and stars were created. Models helped astronomers focus on which observations were relevant. In Otto's opinion, ESP gained understanding much like astronomy 2000 years ago.

Today ESP has hundreds of years of observations from amateurs who had clearly documented their research techniques to sort out non-paranormal experiences. But it was lacking a theory or a formulation to provide a framework that could help establish ESP as a science.

He suggested that the Archaeus Project take the step to produce a conceptual model for ESP. A model would provide guidance and evaluation of each investigation. Now he had them each take the piece of paper he had handed out and hold it in a landscape position. Shown was a block diagram for a model of the human brain and its interaction with the five senses. He had devised it from many detailed descriptions of the brain and simplified it focusing on the major functionality of the sensors and brain. Some areas on the paper had been whited out and something new drawn over them.

Otto admitted that this model of the brain was simplistic, a work in progress, but should be adequate in helping to understand the ESP phenomena. The model was based on the analogy of a digital computer, in other words a machine. Each of the five sensory inputs included a physical body sensor and an external version of the sensor. These combined signals were fed into five Brain Sensory Cortexes labeled olfactory, gustatory, somatic, auditory, and visual. These inputs were processed separately with different discrimination rates ranging from slow to fast. This produced what we perceived as sensed data, defined as Smell, Taste, Touch, Hearing and Vision. Then this sensed data was combined and further processed by what is called thinking, analyzing, comparing, and reacting. The output went into the reporting system (i.e. Speech and Muscle Movements) and into the recording system, otherwise known as memory (i.e. Information Storage System). Otto paused for a moment to make a point as if he didn't believe it himself. He identified several groups involved in memory research and noted that their latest theory proposed that all experiences are memorized, but those tied to an emotional event are recorded with the most detail. Once a memory is stored, it remains. However, memory challenges occur when the conscious brain fails to quickly recall and formulate the memory into a usable thought. Also theorized was that the unconscious brain had access and better recall to this memory because it was not being distracted by immediate response events.

Brain Function Model

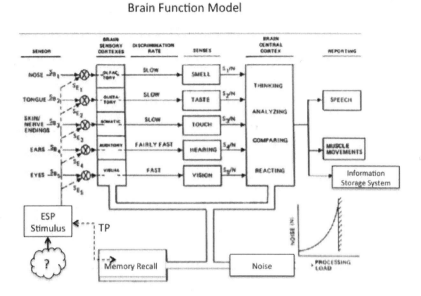

Otto continued with more detail. The physical sensors were listed on the left-hand side of the figure. The output of each sensor was represented as if a single signal S were coming from it, with the subscript B to designate it as coming from the body sensor. And then each signal was numbered so the signal from the nose was designated as SB1, from the tongue as SB2, from nerve endings as SB3, from the ears as SB4, and from the eyes as SB5. Each of these signals went into a summing mechanism that combined these signals with those external to the physical sensors. The external signals were designated with the letter E (SE1, SE2, SE3, SE4, SE5). The corresponding brain sensory cortex, unaware of the source of the signal it had received, processed this combined signal. This modeled the behavior that when people described a supernatural experience they made no

36

distinction between if they had sensed it with their body sensors or through a sixth sense.

Psychic research divided mental phenomena into five categories, four of which aligned with the external sensory signals (SE1, SE3, SE4, SE5) in the model. Clairsentience, labeled as SE1, was the detection of scents when no molecules producing that smell were present. Clairaudience, labeled as SE4, was hearing sounds, such as voices, not emanating from sources in the immediate vicinity of the person experiencing them. Clairvoyance, labeled as SE5, was the ability to 'see clearly' into the past, into the future, and occurrences going on at a distance away from oneself. Psychometry, labeled as SE3, was the ability to derive information from the touch of objects and had been reported being used in criminal investigations. Telepathy, or mind-to-mind communication, was the best-known form of ESP. In this case the ESP Stimulus has a direct connection to the memory and was labeled as TP in the model. The ESP experience of taste involved the fewest reported cases and was represented in the model as SE2.

A function called ESP Stimulus was the origin of the external body sensor signals. It showed two mechanisms for inputs. The first was left open for any number of possibilities that were all uncertain at this time. This was simply represented as a cloud with a question mark in its middle. One theme was that ESP used a new form of energy that had not been detected yet by physics but had often been alluded to in religious stories. The second was from the person's own memory. Many ESP events seemed to be related to a previous memory in the subject.

For most people, the signals from the body sensors are strong, compared with the external sensing. During sleep or meditation, the signals from the body sensors become relatively weak or even shut off, and then the external signals may be strong enough to be detected and processed. Some examples were that ghosts were usually seen in very dark rooms. Mysterious sounds were usually heard in very quiet settings. And an invisible presence was usually felt when in a uniform, ambient environment. This effect of variable sensitivity to the external sensory inputs was included in the model as a Noise generator.

The noise generation was part of the overall connection in the brain between the Sensory Cortexes, Central Cortexes, and Memory Recall. A graph showed the noise as a highly non-linear transfer function. The noise generated into the system was based on processing load and could even hit a wall that would generate nearly an infinite amount of noise at this maximum processing load level.

At this point in Otto's lecture, Dennis couldn't hold back his excitement and offered his experiences related to the model. He said at times he felt he was becoming dumber and slower while in a lively sensory environment, such as a rock concert. He felt as if some of his sensors or his memory were shutting down. At some point he gave up and just let the streaming sensors be the only thing in his mind. The others nodded in agreement as Dennis described his feelings as a bit of an overload.

Otto continued, never realizing he had left some of his audience behind. He said the project should not focus on where each of these 'boxes' might be in the brain but rather focus on defining them as functions. One of the first investigations could try to determine if the ESP Stimulus function acted as a gate to ESP, or as some other even more complex interaction. The model made it possible to accept that the entire ESP experience was contained within the person's brain. The sensory experience would produce an unconscious intuition made as a real memory so the person would respond seriously.

Otto spoke with such authority as he explained the minute details, that the other five in the room gained confidence in the model. They did their own judgment exercise to see if the model fit one of their own beliefs about the connection of the physical sensors and the brain. As Dennis had done earlier, each member of the group shared his or her own thoughts and experiences.

They saw that the model could fit their explanation and supported it. At first Earl thought Otto was being too technical with the group but realized that this detailed discussion was exactly what each member was looking for. The project needed something specific to work from and Otto had provided it.

Otto sat down, as the discussion was now ignited. It did not need a moderator. After they had shared their understanding and confirmation of the model the discussion turned to testing the model. They introduced and talked about various ideas for over an hour, but made no clear progress. Every idea that was presented lacked clarity for consensus. Sometimes the originator of the idea had difficulty verbalizing it. Karen even joked that to understand her ideas they needed a sixth sense. Earl suggested the discussion be shifted toward addressing what questions could be posed about the model. This triggered a much more productive discussion. Dennis started formulating questions as they brainstormed. After another hour he shared this list with the group:

> How to test if ESP events are triggered from memories?
> How to influence the noise level that blocks ESP events?
> Is ESP singular to a few unique individuals?
> Is there training that can enhance a person's ESP abilities?
> When a person undergoes an out-of-body experience, are all the sensory perceptions only through ESP?

Earl then suggested that Dennis, John, Wendell, and Karen each take the lead in facilitating the next five meetings. They should invite experts who studied ESP or people who exhibited paranormal abilities. The group should have the opportunity to question and analyze how the presented subjects fit the brain function model. If preferred, a public lecture could be hosted in the Great Hall of this mansion, and Earl reminded them he would cover any out of pocket costs for these meetings.

Otto added that he would like any visitor to be available to spend a day in his laboratory participating in Electricity of Life experiments. In the years of his own testing on the electromagnetic effects on humans, he had been able to categorize the majority of people to a normal distribution, a bell-shaped curve around the parameter being measured. But quite often there might be one or two subjects that were way off the chart. They were singularities. He felt it was these singularities that could bring new concepts to the understanding of human medicine. And these ESP invited guests were highly likely to be singularities themselves.

Chapter 4 -- Afterlife

Great Hall
West Winds Mansion
2 February 1983

John Hayward stood at what he had planned would be the front of the room, with his back facing the large fireplace in the Great Hall. With seven doors opening into this Hall it was difficult to determine where the front of the room should be. The sun had set over an hour ago and the outside walls looked as black as the inside ones. He was in charge of this meeting, the first experimental meeting of the Archaeus Project, which was to begin at 7:00pm. John had arrived two hours prior to organize the room and review the program with his invited guest.

He was not wearing his priest collar or cloak. In fact, he had dressed like the thousands of other Minnesotans that blended into the winter landscape. He was stepping into a new world tonight and didn't want his past dragged into it. He had convinced himself he could have an open mind about beliefs and facts, even though he had experienced so many things that could only be explained as miracles. He wrestled with the definition of miracles and wanted to open his mind to the possibility of explanations other than spiritual beliefs.

John's path to becoming a priest was unique in that he had come through what some would say was a hole in society that separated good from evil. He grew up on a small dairy farm in central Wisconsin. It was a typical family farm with cows, pigs, chickens, and two workhorses that were more like pets than horsepower. He was content growing up in a rural community. The highlight of each year was a trip to the county fair where crowds, tractor pulls, wild rides, and lots of kids filled him with excitement. When he was a teenager his neighbor, Mrs. Fred Van Hoof, revealed her eight visits with the Blessed Virgin Mary. He was impressed when thousands of people came on the days of the visitations to hear, firsthand, Mrs. Van Hoof's detailed account of the holy occurrence. He had no reason to question her reality and accepted it as he did everything in his life. He believed some things just happened. After high school he attended college in Madison but the University of Wisconsin realized before he did that he was not prepared. It took another year before John realized this himself, and he went to work as carpenter, auto mechanic, and thief. His existence left him empty and he began to drink heavily and use drugs. This only made him lonelier and more isolated. The Vietnam War pushed him further into despair and heavier use of drugs and alcohol. He had avoided being drafted because of some irregularities with his heart. He joined several of the almost daily demonstrations that started on Bascom Hill and marched down State Street to the Wisconsin State Capital. He even broke a few windows along the way to leave a mark of anger. But, like his life, none of this was going anywhere. The city leaders didn't change, the state government didn't care, and even the world didn't notice.

Then one day he was awakened in the alley near a church and his life was transformed. The parish priest became his good friend and encouraged him to seek a better life, a life with purpose. Now, fifteen years later he was a priest. His religious studies paired with his life's experience created a strong interest in the afterlife. The Church provided him a strong foundation, but he was personally drawn to the topic of the afterlife and the role it played in many civilizations throughout history. The most outstanding examples of its importance were evident in the great Pyramids near Cairo and the Valley of the Kings near Luxor. He visited both. He also read as much as he could about the buried armies of Terracotta soldiers near Xian, China, which he hoped to visit someday. These projects were of such grand scale. The importance of the afterlife had dominated some cultures for centuries.

As a result of these studies, John began to distinguish between the soul and the spirit. He researched how others separated these two words, these two concepts. The fact the Archaeus Project was creating a model of the relationship between the brain, the mind, and the spirit aligned with John's personal studies. As a priest, John thought more about the soul than the spirit. The spirit, according to his understanding, was that part of humans that wanted more, that wanted to transcend, and wanted to move into a better place. The soul, the essence of God's creation, wanted an eternal home with God. Closest friends could become soul mates despite their spirits reincarnating in a better place. John felt religious people preferred to believe in the soul's existence in heaven versus a spirit's reincarnation. But of more concern to John was the notion that one's status in the afterlife was a reward or punishment for conduct during life. This created an uncertainty for John because of his own past. He could use the guise of the Archaeus Project to expand a priest's non-religious investigation into the afterlife. He had invited a secular expert on the afterlife. The Archaeus Project was ready to start its first experimental meeting.

Hans Mocker had been born in Germany just before World War II. He could not remember much about the war except he was forever hungry. To this day he never left a single forkful of food on his plate go to waste. He moved to the United States in the early 1950's to complete a PhD in Physics. He currently resided on Long Island developing spacecraft controls for a large aerospace company.

Hans became fascinated with ghost stories while a child growing up in war-ravaged Germany. He pursued science and engineering studies but always kept an open mind about the paranormal. John Hayward had heard about Hans through a relay of discussions between priests. Hans was no longer a religious person but had been raised a Lutheran. He remained open minded about the connections of religion and the paranormal. Certainly, most religions revered stories that he would instead define as paranormal events. His analytical, even what some called cold, approach to data and stories made him an interesting source for discussion of unusual events. He explored as many paranormal events as he could fit into his spare time. Word of his expertise spread and he was often invited to speak to groups interested in the paranormal. He felt these invitations not only allowed him to share his views but also allowed him to collect information about the paranormal experiences of others.

Hans organized his experiences to address three questions. This brought a technical approach to his thought process and knowledge. He preferred the term paranormal experiences but found that people could follow his discussions better if he just referred to them as ghosts. Ghosts were a topic that everyone had an interest in.

The room was arranged with seven rows of chairs facing the fireplace in the Great Hall. The chairs were equally centered in the large room with five on each side of a center aisle. There were about forty people filling the seventy available chairs. Earl and Otto were sitting in the front row to the right of Hans and kept an open chair between them. Hans had found an old wooden cart in the collection of ancient medical devices stored in the mansion and placed it in the front of the room to hold his notes and books.

Attendance at the first open meeting of the Archaeus Project was by invitation only. Each of the six project members invited as many others to attend this meeting that they felt would expand the flow of ideas. Each person came to the meeting knowing little about the planned presentation and was asked to have an open mind and not question the material, but actively observe it. Even though the attendees had come from different areas, some as far away as St. Cloud, there was a commonality to the group. All came in pairs, man and woman. Their clothes looked like they had come out of the same wardrobe, one that had not been updated for years. The styles and colors, clearly from the 1960's, would typically have been worn at church picnics or school basketball games. They were educated and up to date on current events, but had a dull look about them. They had short, straight hair and facial expressions similar to that of someone with

Down's Syndrome. As John introduced Hans Mocker, some indicated by a show of hands they had heard of him, but most had not.

Hans began his presentation by asking three questions:

> When do people experience ghosts?
> How do people experience ghosts?
> Why do people experience ghosts?

He then explained that he used these three questions to organize his results and knowledge of the paranormal occurrences he had either studied or experienced firsthand. There were thousands of books and even more articles written on ghosts by people who called themselves ghost hunters. These writers most often told the stories of others who had had a paranormal experience. Because of his interest in this field he often had the chance to meet with these same people firsthand.

He felt it important to start with what he considered the bottom line, the common denominator, of the experiences he had studied. This was the sheer openness and honesty of the person that had the ghost experience. In re-telling their stories, it was hard for Hans to portray this honesty, yet it was such a powerful feeling when he met these people in person. They had no ulterior motive other than to describe the experience as precisely as they could. It was this unbound disclosure that sustained his passion to understand and define these experiences. Hans spoke in a clear, self-assured voice avoiding emphasis of any words. He said his goal was to recreate, as accurately as possible, the stories and settings that each person shared with him. There was hushed silence as Hans continued.

Hans reiterated the first question, "When do people experience ghosts?" In his findings over the past twenty years he said about eighty percent of the time people report seeing ghosts when they are near death or had a near-death experience.

There is the story of Sally in Kansas, who was dying from cancer. She had a vision of her deceased sister, who beckoned Sally to join her. Sally vividly described her sister hovering in beautiful white clothes. Her sister spoke only a few words that were most meaningful only to Sally. She talked about the day they swam in the lake, when they both realized how much they loved each other. No matter where their lives took them they would always look after each other. Now it was her sister's devotion that would help Sally transition to her next life, her afterlife.

At the time of this visit, several of Sally's family and close friends were gathered around her in her own bedroom where she had chosen to spend her final days. However, this move seemed to cause Sally to linger in pain while her family looked on, torn between keeping her with them or letting her go. Sally slept most of the time due to the heavy pain medications, but when Sally's sister appeared to her, Sally was alert and focused on a dark corner of the wall and ceiling. Others, who were surrounding Sally as she lay in her bed, looked themselves to see what Sally was fixated on. They could not see, feel, or hear anything special that captured Sally's attention so completely.

Minutes later, Sally blinked her eyes several times and described to those at her bedside that she first heard and then saw her deceased sister in the upper corner of the room. How bright it was! How beautiful her sister looked! How assured her sister was for them to be together again. Forever. Family reported there was a mist in Sally's eyes and her expression was blissful. In that moment she closed her eyes, reclined her head and shoulders, and exhaled her last breath. She died in ethereal happiness.

Hans then described five other stories where people reported hearing and/or seeing deceased friends or relatives. Not all involved the person being called to the afterlife, as Hans had described it. Some had been told it was not their time and they must remain with the living to finish their work. In his studies all the people who had experienced a ghost where they were told to return to the living were not dying from incurable diseases but were involved in tragic accidents.

There was the case of George, whom Hans had interviewed at great length after he was involved in a serious automobile accident. He was driving with his wife in the passenger seat and their two young children in the back seat. The car was T-boned on the passenger side by a semi-truck when George had missed a red light. The car was upended and tossed fifty yards off the road. He remembered the crash and the car coming to a rest on its side. He blacked out for some time and was awakened by his wife. She was beautifully dressed and hovering outside the car in a sea of white light. She spoke softly to George insisting that he stay and not accompany her. His life was not complete and their children needed him. As she faded away he remembered first responders using a jaws-of-life to open the car and pull him back to life. It was only this vision and the heart-felt appeal from his wife that allowed him to carry on and make a new life for their children. George said he told this story as praise to his wife for the strength she showed in extreme anguish and tragedy.

If the room was quiet before, now not a thing was moving. The only sound one might hear was the west wind trying to sneak through the large doors on the south side of this Great Hall, which was a sound that was best not heard. Otto shifted in his chair and raised his hand to ask a question. Hans waved his right hand in a subtle motion indicating he wanted to continue without interruption. Hans summarized that these stories were intended to illustrate the most common scenarios when people experienced ghosts.

Hans stated his second question, "How do people experience ghosts?" Again, after his review of hundreds of documents and the shared firsthand experiences, he reasoned that ghost sightings were nearly the same as an extrasensory perception, or ESP. His brief answer to his second question was that people experienced 'ghosts' the same as anything else, with their five senses. Based on this, he understood why the Archaeus Project had invited him to be present at this meeting.

Many people reported seeing a dead relative, hearing a favorite school teacher, feeling the presence of a brother who had died unjustly, and even smelling the pipe tobacco of a grandfather or tasting mother's prize-winning jam. As best as Hans could determine, all these experiences were personal self-awareness. If others were nearby they may also be having an ESP experience but it would differ from person to person. This indicated to Hans that ghosts were experienced by a person's brain through their sensory channels, but not through their body senses.

Hans mentioned that Minneapolis was a notable place where people experienced ghosts. For over 90 years people working in, visiting, or living in (if you call being in a jail cell living) Minneapolis City Hall have experienced a ghost. In fact, this ghost, they say, is the spirit of Mr. John Moshik. The city hall was first built in 1888 as a huge glorious structure, reminding one of a Florentine castle, complete with two towers. The north tower had a large clock on one side and produced bell music for downtowners to enjoy. The South Tower, part of the jail system, was a place for men on trial or awaiting executions. In 1898 John Moshik was hung to death in this tower after being convicted of murdering another man during a robbery. This hanging did not go well because of mechanical issues. It was reported that John hung in agony for what seemed like an eternity with his screams penetrating every room and hallway in the building. To this day workers, inmates, and visitors say they've heard footsteps, felt icy breezes, and even spotted John Moshik. Partly because of the botched hanging of John Moshik, no further executions had taken place in the City Hall.

Now Hans pointed to five people who were sitting to his left, three rows back. They filled the row on that side of the room. He said that each of them worked at the Minneapolis City Hall and had experienced the spirit of John Moshik. He asked each to re-tell his or her experience.

Andy, on the outside of the row, stood first. Most visible to the others in the room was his large parka that he kept zipped as the hood flopped behind him. He was about 50 years old and had plenty of graying brown hair. He worked as a guard at the detention center which was now on the fourth and fifth floors of a remodeled part of the City Hall. Several nights when he was doing his rounds he had heard footsteps but when he rushed to where the sound was coming from there was no one there. Then the sound might start again from where he had just been. He had stopped chasing these sounds since it seemed like someone was just having fun with him. Actually he was convinced that it was John Moshik having fun with him and trying to scare him into retirement.

Cindy stood next and also had on her long winter coat over her office attire. She worked in one of the City Hall offices organizing court cases. She had seen the ghost of John Moshik two times. It was always when the room was empty and John seemed to be trying to distract her from preparing all the evidence in an orderly manner.

Next was Sue and she had also seen the ghost of John Moshik three times. It was at the most inopportune time, when she was hurrying to finish her work at night. She said she was usually behind schedule because they were understaffed. Bob, who smoked and had to take breaks every couple hours outdoors because of Minnesota's Clean Indoor Air Laws, stood next and said he felt breezes in hallways and the courtrooms with no doors or windows open.

Lastly, Bill stood wearing all his winter clothes including his flannel styled baseball cap. Bill was usually in City Hall as a court appointed lawyer. He had both seen Moshik and felt his presence at numerous times. There was no other explanation for what he had experienced. He listed a dozen other possibilities but ruled them out one by one from his own investigations.

Hans thanked the group and said these five fit the description and spoke with conviction and honesty about his or her experience. It was clear none of this was rehearsed and that each believed there was a physical spirit in the City Hall. They each responded they did not have any physical evidence to share, but their experience was very real, just like any other memory they had.

Hans reiterated his third question, "Why do people experience ghosts?" He said there were two possibilities. Either dead people wanted to communicate with the living or the living wanted to communicate with dead people. In summary, it was all about the afterlife. ESP was the communication method with the afterlife. Did the existence of ESP prove the afterlife? Did the existence of the afterlife prove ESP? Hans said he didn't have any way to answer these questions. He was a paranormal researcher, perhaps an expert, but he was limited to observations. He was very good at recognizing holes in stories where facts would conflict with each other. But even after allowing for this there were still thousands of stories and more coming each day. Ghost stories came from every part of the globe and from every culture, perhaps because the desire for an afterlife was so basic to the human id.

Hans now pointed to his host, John, and said that maybe it was better for his profession to figure it out. John quickly stood and described to those that did not know him that he was in fact a priest and curious about how science could further the beliefs of religion. Then John asked Hans if he thought there were any ghosts in this mansion. Hans said he had developed a method to reveal ghosts. He had learned that if one spoke disrespectfully about the dead most closely associated with the building or site, you could almost count on a reaction. Hans asked Earl if he knew anything about the people who first lived in the mansion. Earl replied that William Goodfellow built the mansion over fifty years ago. It seemed he was a member of the Rosicrucian Fraternity, which explained the strange architecture in each room that played to both a Gothic and Pagan theme. He built the place to be a safe place for his future family but sadly he never married. It seemed that a housekeeper, named Ruby, was the primary resident in the history of the mansion.

Hans changed the story he had just heard from Earl, putting a negative spin on William Goodfellow. He suggested the reason Goodfellow never married was because as a Rosicrucian, he was more extreme in his beliefs about mankind and that he was hiding an affair. He also suggested that maybe Ruby really didn't like children or families and just wanted the mansion to herself. Hans saw his audience become quite uncomfortable with this defamation of people they didn't know.

At that moment, the sound of loud pinging noises surrounded the north and east walls of the Great Hall. It was like a whip had raced just outside both walls in a speed faster than any person could run. Hans held up a red and white ball between his thumb and forefinger in his left hand. He said that being in Minnesota everyone would recognize the item. It was a small fishing bobber that was half red and half white. He explained that fish could not easily see the red half of the bobber underwater but the fishermen could easily see the white half above the water. Actually, any dark color would work for disguising the bobber from the fish but red was used because it was the cheapest pigment. Hans said he used them in his paranormal research because they were inexpensive and extremely lightweight. They also were not very stable and when placed on a hard surface it took the slightest amount of energy for them to start to move. Hans explained he had carefully placed many of them in the hallways outside this Great Hall. They were on top of lights, shelves, railings, and chairs. The slightest motion or temperature change could cause air movement to disturb and displace them. What he guessed just happened was that something going the speed of sound passed through the two halls outside this room and caused his carefully placed fishing bobbers to fall. The noise seemed to start directly behind everyone and then move to the outside right hand wall stopping in the entrance hallway near the stairway heading to the second floor. Someone in the back of the room opened the door and announced they could see several bobbers scattered around the floor.

Hans calmed everyone as he brought attention back to his presentation. He said he had learned that, if spoken poorly of, a ghost could sometimes be stirred into action. Outrageous lies aroused them to show themselves when their own reputation was being tarnished. Hans concluded his talk by suggesting this mansion did indeed have its own ghost who was not Mr. Goodfellow but most likely the housekeeper Ruby. He wished the Archaeus Project good luck in its quest to define ESP and to hopefully not have much interference from the presence of Ruby.

The audience gave Hans a loud round of applause. John jumped up and shook Hans's hand followed by Earl and Otto. Earl asked Otto if Hans should visit Otto's laboratory before he flew back to New York. Otto replied that his laboratory did not have any ghosts and Hans might be happier visiting a cemetery.

As he was leaving Otto inspected the fallen bobbers in the hallway. Hans, in a panic, didn't want any help in picking up the bobbers. He said he needed to inspect each of them carefully. Otto noticed small long hairs that seemed out of place near some of the bobbers. As Otto walked out the front door he also wondered where the sonic boom was if the object that caused the bobbers to fall was moving at the speed of sound.

Later that night John was sitting in his own living room. He was alone in the dark. His mind was filled with scattered thoughts, all centered on this concept of the afterlife. The mystery of Mrs. Van Hoof came back to him. How could she be the only one with thousands of others present to see and converse with the Blessed Virgin Mary, who had been dead for nearly 2000 years? The main message, demanding greater devotion to prayer and penance, was relevant to the times. Also, as Hans had alluded, there were thousands of first person stories of communications with the dead, or more gently stated, their ghosts. The concept of a ghost did not seem to fit the same concept of a soul. Hans had expanded John's knowledge of afterlife experiences that would further influence his future understanding of life and death. He had to open his mind further to all the mysteries of life. His religious background was maybe enough for a student but not complete enough for a teacher. He hoped this Archaeus Project would establish a better understanding of the occult. He pinched himself as he considered this pagan thought.

Chapter 5 -- The Astral Plane

Great Hall
West Winds Mansion
2 March 1983

Dennis Shilling stood at the front of the Great Hall. This Wednesday night was again fraught with bad weather. March had come in like a lion. Only a few of the seventy chairs were occupied. He announced he would delay the start of this meeting of the Archaeus Project by thirty minutes. He walked to the south wall looking out one of the windows. The blowing snow blocked any view of the garden, street, or even streetlights. The black space and flying snow pellets allowed just a few inches of visibility from the light of the room. He decided he would offer some discussion before he had the invited speaker begin.

He took a chair from the front row and turned it so he could sit and face those in attendance. He talked about his part-time role of helping to establish and organize the Bakken Library, which was the main use of this West Winds Mansion. Earl Bakken had first asked him to seek out and collect historical medical instruments and devices. These were difficult to find, as there was no buying and selling marketplace for such things. However, he was able to purchase hundreds, even thousands, of old books written on medicine and the human body. Anyone interested could come explore, read, borrow, and research from any of the materials available. Dennis said that recently they were having some luck in obtaining medical treatment artifacts. He stood up and walked toward the door to the right of the large fireplace that opened into the study. He was thinking about showing the electromagnetic penis straightener when Earl entered through the side door from the hallway. Almost running into Earl, Dennis quickly turned around and straightened the chair back into its proper formation. Everyone waited for Earl to offer his thoughts.

By now there were about forty people in attendance and Earl suggested that Dennis start the meeting. Dennis introduced himself again and said he had selected the topic, The Fourth Dimension, for tonight's meeting. As they had learned in last month's meeting, Extrasensory Perception, ESP, was an explanation, a mechanism for communicating with the afterlife, whatever the afterlife was. He preferred to think of the space beyond our physical world as the fourth dimension. It was also named the Astral Plane. Plato first devised the astral concept stating that the human was composed of 'a mortal body, an immortal soul, and an intermediate spirit'. The Astral body had come to be understood as the median between the rational soul and the physical body. The term Astral was derived from the Greek word meaning 'related to the stars', and was originally used to describe the living space of their gods.

Earl was sitting in the front row beside Otto as they had positioned themselves for the first meeting. Earl gave Dennis a propeller sign with his right index finger to move on. Dennis then quickly introduced Nick Zemma who had arrived yesterday by train from North Dakota. Nick started by saying he felt right at home with the clipper-like snowstorm going on outside. For years Nick had been invited by various groups from around the world to talk about his understanding of sensing and to interview people who had what they called 'out-of-body' experiences. Nick planned to describe his interest in brain function and hoped the Archaeus Project team could relate it to ESP, but he didn't know how that might really be done. He then launched into his description of the fourth dimension.

Nick continued by defining some terminology. Our three-dimensional world, the space we lived in each day, was defined as the Physical Plane. Nick paused for a moment. He commented that he preferred to visualize dimensions like Dennis, but most people could not follow the visualization of a fourth dimension. The three dimensions where we know the soul exists were connected to the fourth dimension through the spirit. Also, since the introduction of the phrase 'Black Hole' twenty years ago, astronomers had been expanding theories on multi-dimensional universes. They reduced the concept of a fourth dimension into a mathematical abstraction that could represent any number of dimensions. Much earlier paranormal literature adopted the terminology of planes. The world around us, the three dimensions we move in, was called the Physical Plane. The Astral Plane was what was beyond the Physical Plane. A Plane of Forces connected these two planes. The Forces of Nature were not perceptible to the physical eye, except as manifested through matter. One cannot see electricity but might receive a shock and realize its reality. One cannot see the force of gravity, but becomes painfully aware of its reality when falling.

Nick then looked at Dennis and said he agreed with Dennis's origin of the word Astral as 'related to the stars'. He then added that over the years Christians had preferred to label this Astral Plane as 'heaven' and pagans had labeled it as 'the land of ghosts'. In either case the Astral Plane was believed inhabited by beings of an ethereal nature or simply disembodied spirits.

In the Astral Plane each of the physical senses had its astral counterpart. Thus, when people experienced the Astral Plane they reported the power of seeing, hearing, feeling, smelling, and tasting. Nick took a small step sideways and said that maybe ESP was this Astral Plane. Nick had investigated many people who accidentally, or through careful training, developed the power of Astral Vision where the scenes of the Astral Plane were perceived just as clearly as those of the Physical Plane. A well-trained Occultist could shift from one set of senses to the other on command.

Some paranormal researches focused on a segment of ESP they called Out-Of-Body-Experiences (OOBE). OOBE's were the case studies of the Astral Plane. There were detailed descriptions of OOBE's that had been documented throughout recorded history. They tended to be once-in-a-lifetime experiences that occurred under special circumstances. The description from the person having the experience was described as correct and more accurate than what could be expected by pure coincidence. And, as opposed to ghost encounters, the experience was usually extremely joyful. OOBEs could only occur under a limited set of circumstances. These were while asleep, while undergoing anesthesia, at a time of an accident, at a time of intense physical pain, and in deep meditation.

A recent story that illustrated the experience of the Astral Plane involved a butcher Nick had recently met with in Canada. Nick had visited the butcher after the butcher had written Nick a letter stating he had had an unusual experience and didn't understand it. The butcher welcomed Nick's visit and hoped to gain some sanity for his own good. Nick said he was telling this story of the Canadian butcher but there were hundreds of books filled with thousands of firsthand accounts of Astral Plane experiences similar to this.

Nick then told this story as told to him by the butcher:

> *A month ago, I was not feeling well and needed to rest in the middle of the day. This was an unusual occurrence and I couldn't remember a time I did this before. I had to get some rest and left my young shopkeeper in charge of the butcher shop while my wife was out with a friend. I went to my bedroom above the store and collapsed into bed. At some point I awoke and was able to see through the floor to the checkout counter below. I first felt a draft from the meat locker door being open. I could see the young clerk eating a hot dog and offering one to his friend who had stopped by. I watched them and especially noticed they didn't pay for the meat. I didn't have any specific emotion about it and just observed it. Then that vision faded and suddenly I was peering into a small bakery that was several blocks away. I could smell coffee brewing. I recognized the place as I often went there for a coffee and a small cake when my wife and I had something to celebrate. Now I watched from above as my wife entered the small shop with her friend and they both ordered a large piece of cake and*

followed it with a large piece of chocolate. Here I felt some emotion, maybe anger, because my wife was always making us be conservative in the availability of sweets. This vision then faded away and later I awoke to find myself still in my bed. I didn't feel any specific emotion and proceeded to return to my store below. I told the young clerk he could leave but before he left he should pay for the two hot dogs he had with his friend. The clerk was quite shocked by this comment. The butcher said he could see the young man wanted to ask how he knew but instead sheepishly put the money in the cash register and left. Later my wife returned from her afternoon out and immediately said there would only be a light dinner tonight because we must watch our weight. I replied I had expected so since she had just eaten such a large piece of German chocolate cake, cut into a square and topped with whipped cream, specifically requested by her. I even said I could smell the hint of cherry added to the cake. It was a serving and a half followed by two large pieces of the cream-filled chocolates the Polish bakery had just started to carry. My wife looked at me with astonishment and asked how I knew so much about her afternoon snack. It was more detail than even she could remember. I didn't answer her. I couldn't answer her.

Later I grew to resent sharing what I had seen from this Astral Plane experience. His wife relayed the story around town and everyone was convinced I was spying or had hired someone to spy on her. Our relationship, an example of faithfulness and devotion, began to have moments of awkwardness. Silence was growing between us. But I continued to stick to my story and my paranormal experience. It was the truth. Finally, I sought you out to learn why, and maybe how, but mostly why did I have this experience.

Nick then broke out of his role-playing mental state and continued the story from his own perspective. He had assured the butcher that what he had experienced had a purpose and there were thousands of similar experiences that had been recorded since the beginning of time. Maybe not exactly about hot dogs and chocolate cake, but about things which were equally important to those involved. Nick assured the butcher, and then his wife, that this type of Astral Plane Experience was a sign of caring and definitely not a sign of mistrust. Many people had difficulty in expressing their feelings openly and, in these instances, his or her unconscious created avenues for expressing them. It would seem that by exposing these pent-up anxieties a future anger was being defused, and a more open relationship resulted.

The Archaeus Project audience sat mesmerized by this story and how easily Nick portrayed it. It was told with such honesty that everyone was ready for the next part of Nick's talk. No one tried to interrupt Nick with questions. This was an unusual state for the Archaeus Project that only lasted till Otto came out of his trance. Otto startled the quiet by firing off a list of thoughts:

Why does he focus on people eating?

Is it related to him being a butcher, a provider of food?

Is it what he desired for himself?

Why does he accuse only the people close to him?

Would he want to do that same thing, but he thinks it is wrong?

Is it his moral compass testing its purpose, its importance?

Nick was taken back by such a list of rapid-fire questions. He paused for several minutes with the audience now focused on him to respond. People wanted to know why. He had given up on the why long ago. Each case he studied was unique, a singular event, and it was difficult to see a common thread between them. He was content to document what was observed. It was difficult enough to capture what really happened, at least as best that could be pieced together between all that were involved. Finally, he replied softly that they, all of them in the audience, had invited him to tell them WHAT he knew. Nick continued that he now wanted to invite The Archaeus Project to tell him WHY these paranormal experiences happened.

Nick proceeded to describe Astral Plane visits associated with near death experiences. Otto was anxious to hear these descriptions. These out-of-body experiences seemed to occur in similar situations as people being visited by ghosts. When in a near-death situation a person could see his body laid out before him, completely dead or nearly dead, separate from himself. Sometimes the OOBE individual recalled moving toward a bright light that signified an afterlife. While it was true that experiences like these were difficult to take seriously or dismissed as mere hallucinations, it had been proven that both, the sensation of movement and 'seeing' a bright light could be triggered during an unconscious state. This could be described as imagined, but Nick had interviewed hundreds of people who experienced it, and it was very real to them. In his opinion, when people just imagined things they didn't describe them with the same seriousness as an out-of-body-experience.

Nick paused for a long time as he looked around the room. It appeared he was judging the audience about how he should proceed. He walked away and then back to the podium and stood directly behind it with his hands hidden inside it. He said that maybe some had noticed his behavior of walking back and forth along the front of the room and occasionally returning to the podium. He paused again and then when the room was dead quiet, except for the wind that swirled through the trees outside, he said he had been doing some experiments since he had started to talk. He was doing experiments on the Archaeus Project audience themselves.

In considering these many experiences, he tried to devise some physical explanation for these Astral Plane visits. He said he had given up the idea to try controlled experiments on these events. He had come to the conclusion that each of these astral experiences was a singularity. The scientific method couldn't be applied to singularities. It needed situations that had repeatability, reversibility, or better yet parameters that could be measured statistically. In an instant Otto jumped up from his chair and concurred with Nick's assessment. Otto then summarized that in his own biophysics research, which had spanned four decades, he had come across several happenings or measurements that could only be defined as singularities. They typically were off the chart of a normal distribution of that parameter. Many times there were not enough people on the planet to statistically think that such a value was probable. Yet there it was, coming from a subject that was in his own laboratory.

Nick patiently stood and listened to Otto as he spoke in his excited, arm waving style. Eventually Otto nodded to Nick and sat back into his chair. Nick continued that one approach to connect the Astral Plane experiences and the human mind was to consider electromagnetic waves. These were the energy carrying part of the electromagnetic force. The electromagnetic force was the most prevalent to humans. It was the force that allowed one to chew and digest food, smell flowers, jump into the air, kiss someone, and turn your back on others. Two of the other four forces of nature, the Strong and Weak Forces, were only evident at atomic dimensions and the fourth force, Gravity, was most evident at galactic dimensions. These other three forces were certainly critical in making the entire environment what it was, but were not forces humans could directly interact with.

Nick explained that when he first studied electromagnetics, or EM for short, he quickly focused on the EM waves associated with the brain. There were four categories—beta, alpha, theta, and delta. Each represented a different set of EM frequencies and could be associated with different states of the human mind. The human mind was thought to have three operational states or capabilities—conscious, preconscious, and unconscious. The different patterns of the emanating EM waves were discovered to be associated with the operational states of the mind.

1. Beta waves ranged in frequency from 14 to 40 hertz and were evident when the brain was awake, conscious. Thinking and doing actions like even blinking an eye caused large changes in these emanating waves.

2. Alpha waves ranged from 7.5 to 14 hertz and were present when the eyes were closed and one was slipping into a daydream or during light meditation. These waves acted as a gateway to the unconscious mind and were associated with a preconscious awareness.

3. Theta waves ranged from 4 to 7.5 hertz and were present when the brain was sleeping. Theta brain waves were present during the all-important REM dream state and emanated from the realm of the unconscious mind.

4. Delta waves ranged from 0.5 to 4 hertz and were present when the brain was in deep, dreamless sleep and in very deep, transcendental meditation where awareness was fully detached. Delta waves were the gateway to the universal mind and the collective unconscious, where information received was otherwise unavailable at the conscious level.

Nick then passed around sheets of paper illustrating each of these brain wave patterns. These waves were detected with electromagnetic field sensors located close to the brain. The graphs did not resemble smooth, uniform patterns. These looked more like some scribbling and scratching as if to cover up a mistake hidden below the wave pattern. Certainly, the members of the Archaeus Project had known of brain waves and their definition. Otto, with his wife Viola, had been doing years of research on effects of electromagnetic waves on the body. They had measured these EM brain waves, particularly researching the Alpha brain waves. One of their breakthrough discoveries was the ability to synchronize a person's breathing and heart rate by monitoring the Alpha waves. This was one of Otto's many medical discoveries that had not yet found a value in medical treatment.

Nick asked the audience to reconsider the case of the Canadian butcher. The butcher had not left his upstairs bedroom otherwise his young worker would have seen him. But he was still able to experience with such confidence not one but two activities that he felt he had witnessed in person. He shocked both his worker and his wife with vivid and accurate descriptions of things they each were hiding from him. We can surmise that during his rest, either in a dream-like or meditation state these images came to him. The presence of the alpha, theta, and delta brain waves prove that the brain is very active when the body is inactive. During this state the body sensors are also inactive, yet the butcher's remembrance of the two experiences included colors, smells, and feelings of temperature. Nick explained that the brain must have created this extrasensory perception and fed it through all the sensory channels such that the memory had recorded them the same as if in a fully alert state. In Nick's opinion the brain had a tremendous capability that wasn't understood. Scientifically understood or not, it could provide experiences that changed many people's lives. The confusing aspect of this was that we could not distinguish between what we had recorded with our body sensor channels and what had been recorded with our extrasensory channels. Another interesting human trait was what was called a 'slip of the tongue'. In Nick's opinion the unconscious mind had prepared a deeply held truth that spontaneously entered the conscious mind and was rapidly streamed into speech. Nick said he had given up on trying to hide anything anymore because this so often happened to him.

Otto slowly woke from a trance like state. He noticed the rest of the Archaeus Project group opening their eyes and slightly stretching as if coming out of a daydream. Their speaker, Nick, was gone. Sitting on the podium was a small electronic box and a note. Dennis walked to the podium and held up the cigarette pack sized electronic box. It had a blinking red light, a small switch and two rotary knobs on its face. Dennis read the note aloud:

> *Thank you for the invitation. Approximately thirty minutes ago I left the room, and the building. I started this device*

some time before that. I built this device to stimulate alpha waves. You can turn it off now. I leave it for you to do some of your own experiments. One of my favorites is to take it to restaurants. I have found by making the adjustments in frequency and amplitude the device can encourage people to enter a daydream state. As we know this encourages and heightens your imagination, visualization, memory, learning and concentration. I suspect that since I turned this device on, each of you has been able to complete my presentation using your own ESP with the unconscious mind filling in the wanted knowledge. Maybe each of you can share your exact memory of the last part of my presentation and you may be surprised how similar they are. Oh, there will always be differences and as we know even eyewitnesses of the same situation often differ in exactly how they remember the details. Maybe some of you just daydreamed about other topics. But I suspect that those closest to the Project, having similar objectives and preparations for this lecture, have created very similar memories. If you compile the same facts similar computers will arrive at the same results. Enjoy the rest of your waking night. It may seem I am running out to avoid more WHY questions. I am not avoiding them so much as I have yet no answers for them. Hopefully my observations can provide some basis for your pursuit of the deep meaning of the Astral Plane experiences. I am off to catch the 10pm train to North Dakota.

Chapter 6 -- Spontaneous Images

Great Hall
West Winds Mansion
6 April 1983

Dr. Karen Ness opened this third public meeting of the Archaeus Project by waiting until ten minutes past the announced starting time before entering the room from the back and slowing walking to the front. She talked then stepped, and talked and stepped again, bringing importance to each line of her opening presentation. Here was what she said:

> *The form on an Extrasensory Perception experience is distinct from its type.*
> *(step)*
> *In type it may be related to one of the phenomenon in the model.*
> *(step)*
> *Clairsentience connected with smelling.*
> *(step)*
> *Clairaudience connected with hearing.*
> *(step)*
> *Clairvoyance connected with seeing.*
> *(step)*
> *Psychometry connected with touching.*
> *(step)*
> *and Telepathy as spontaneous memories.*
> *(step)*
> *The type shows which senses were primarily involved.*
> *(step)*
> *(pause)*
> *(step)*
> *There are equally five forms for ESP.*
> *(step)*
> *Intuition.*
> *(step)*

Realistic Dreams.
(step)
Déjà Vu.
(step)
Unrealistic Dreams.
(step)
Hallucinations.
(step)
The form shows the psychological manner in which the information emerged into the consciousness.
(step)

Karen had now reached the front of the room. She turned 180 degrees to face the audience and continued.

All these forms can be grouped under the phenomenon of spontaneous images.

Karen then proceeded to introduce herself before introducing this month's invited guest speakers. She was a doctor at the Minneapolis Children's Hospital specializing in Behavioral Pediatrics. She was an expert in the use of biofeedback in place of medication to treat various physiological problems. She had reported hundreds of cases where therapy, not chemicals, was able to repair a child's problem in some bodily control, such as urinating. These treatments took longer than surgically implanting devices and as a result were more expensive. But she felt this ability to tap the hidden parts of the mind offered longer lasting and vastly new possibilities for future medical practices.

Karen spent a minute offering insight into her own motivations for researching ESP. Just seven years ago the first study on what became known as Psychophysiological Controls was completed. The study included 60 children, aged 7 to 9 years, half of whom were learning-disabled. The experimenters gave verbal instructions to children about controlling their fingertip temperatures, including suggestions about warmth and coolness. Temperature sensors were attached to the index finger of each child. When children successfully changed fingertip temperatures, a biofeedback mechanism translated the amount of temperature change to move a toy locomotive around an oval track. About thirty percent of the children were successful at this. And, interestingly, the learning-disabled children in the group achieved a greater degree of control than did normal children.

Since then her own research into Psychophysiology led her to more deeply understand the human mental (psyche) and physical (physiological) processes. She realized she needed to become an expert in the interactions between mind and body. She also realized that her own training was limited and the knowledge and ability of psychologists, biochemists, engineers, to name a few, were also needed to understand these interactions. Karen felt that Psychophysiology must be part of the Archaeus Project. First, the Archaeus Project group had the diverse backgrounds needed to study the mind and body and second, the results of such a study aligned with the project's ultimate goal: Providing innovative medicine by recognizing the intricate connection between mind and body, thus doubling the quality of life at half the cost. A psychophysiological disorder was characterized by physical symptoms that were partly induced by emotional factors, such as anxiety, stress, and fear. These manifested themselves in psychosomatic ailments, most notably migraine headaches and hyperactivity. Treating these ailments often did little to address the psychophysiological disorder.

A recent finding that her patients reported spontaneous images when the desired bodily control was triggered excited Karen. Analysis of the spontaneous images led her to speculate that images were vibrating forces of specific amplitudes and strengths, which could cross synapses in electrical form and convert into a neurotransmitter, which then initiated a cascade of events culminating in a physiologic response. How many vibrations were there? How many neurotransmitters? How could children demonstrate such precise control over this process? Why did these tasks seem more difficult and less successful when the subject "tried hard"? Ultimately, the human form was a composite of vibrations in atoms, molecules, and other intellectual abstractions, which had assumed a cellular form but were nonetheless composed of myriads of vibrations. Negative thoughts translated to observable negative physiologic responses. In the experimental setting, she had demonstrated that planned, positive images also translated into positive physiologic responses previously believed to be involuntary. She wondered how long the health-care industry could go on using crude diagnostic and treatment methods that did not consider image-generated effects on physiology, drug absorption, drug transport, laboratory values, nutrition, cardiovascular responses, respiratory responses, immunologic responses, and neuromuscular responses. Someday the relationships between patients' images and their responses to treatment would be fully understood. And then patient care would have to consider the relationship of the therapists' own images and their decision-making. Karen knew exactly what she was talking about, few others in the room did.

Karen then looked around the audience nodding to Otto and Earl who sat in their usual spots at the front of the room. She had invited two guests to capture the current understanding of spontaneous images. The man was quite familiar with Minnesota, as he had been a University of Minnesota law professor for most of his career. After retiring he and his then recent wife had moved to Colorado where they developed and taught courses on parapsychology. They retired a second time and now lived in Wisconsin. Mary Jo and Walter Uphoff were involved in full-time pursuit of the paranormal. In the past ten years they had published two books that were widely read throughout the paranormal community. On cue, Dr. and Mrs. Walter Uphoff entered the Great Hall behind Karen from the door to the right of the fireplace. Otto immediately jumped up to greet them, as did Earl. The audience gave a respectful applause of welcome.

Walter and Mary Jo walked behind the podium standing side by side. They looked like they were cut from the same mold. Both had bleach white hair and matching horn-rimmed glasses. Walter was a few inches taller than Mary Jo, she was a few inches broader at the waist. Both wore clothes of the 1960's style. Mary Jo held up two books and said they would be willing to autograph any that were purchased at the end of the meeting. In 1975 Mary Jo and Walter had summarized much of their knowledge in a book titled New Psychic Frontiers: Your Key to New Worlds and in 1980 documented their own advances in knowledge of the paranormal in a book titled Mind Over Matter. Walter nodded to his wife who then sat in a chair in front of the podium.

Walter began the presentation by telling the story of twin boys he knew when he was growing up. They were identical twins that were neighbors to his family's small dairy farm in central Wisconsin. One twin was left-handed and the other right-handed. It was nearly impossible to tell these two apart. Walter said his father often threw a small object, like an old bolt, for the boy to quickly catch to identify which twin it was. However, he still did not get the name right because his father could not remember which name went with the right- or left- handedness. The left-handed boy studied mathematics in college, the right-handed one studied physics. The left-handed one played first base in baseball and the right-handed one played shortstop. If one chose a red plaid shirt to wear to school the other chose a blue version of the same design. Often, they would only discover this when they got to school and took off their overcoats. As they got older they didn't want to dress like twins so they had to make a conscious effort to wear non-matching clothes each day. When they forgot to check they often ended up matching for the day, much to the amusement of their friends.

There were two very unusual characteristics about these twins. First, they were always in agreement. If one wanted to date a certain girl the other was in total agreement with the choice. If one wanted to do certain chores on the farm the other agreed to do the others, even if the work was not fairly divided. If one wanted to borrow the family car, the other had no need for it. There was never a conflict. Not just that a conflict was easily resolved but they never faced a point of conflict. The second characteristic was that they could communicate at a pace others could not follow. Whenever they were in a group discussion about a lively topic, the two would eventually dominate the conversation, as if no one else was there. The topic didn't matter. It could be – who was the best professional baseball team; which teacher was the best; why did they lose the basketball game; or even what was the best way to build a tree house. Walter experienced these run-away conversations many times. Most people just walked away saying that the twins were speaking in their own language again. Walter stayed and tried to figure out what was happening. At first, he felt that one twin was completing the sentence for the other. But as he listened more carefully this was not the case. One knew the rest of the sentence and just started on the next. And when this sentence was partially spoken the other twin would start on yet the next sentence.

For thirty years Walter had thought about this method of conversation. Before Mary Jo and he focused their fulltime attention on the paranormal, he thought these twins were an excellent example of the finest telepathy possible. A new energy form was the source of this amazing level of communication. However, recently he had come to conclude that there was no other energy being transmitted that anyone present would not also receive. A better explanation he felt was that these two were communicating with images being generated in the mind. Because of their identical makeup and experiences, they were able to process nearly identical results. Once the image was formed the other could start with words, expressions, and hand gestures that launched the next image processing cycle. In this way, the two completed an entire discussion while the rest were left behind. At some point those still present would not know what was being debated or even concluded until one of the twins reviewed the conversation in complete sentences for the benefit of the others still listening.

Walter paused, staring at his wife, Mary Jo. He started again slowly saying that one did not need to be a twin to have this communication ability. He had met Mary Jo later in life, just ten years ago. From the beginning, especially when they were dating, they each had déjà vu moments almost daily. When one said they should go to a movie, the other reacted as though he had thought the same thing earlier. When one would call on the phone spontaneously, the other would answer the phone before it rang, not even realizing it. Walter asked the audience if they ever had these moments, and nearly everyone nodded and looked at others nearby.

Mary Jo stood up and faced the group. She wanted to offer her chemistry viewpoint. She said that when one was young it seemed more like being an ionized atom and as a result opposites attracted each other. A good example was positive charged sodium atom and negatively charged chloride atom combine to make salt, an essential part of human life. But if one has the chance to re-mate later in life the relationship might be more like two molecules. As life evolves a person takes on other elements that are engrained in the individual. This baggage is almost inseparable from the self and can be imagined more as a molecule. Molecules were the most stable when around others that were a near exact match to it. Maybe they didn't have as much spark compared to atoms, but for sure the romance was more comforting in the long run. Walter appeared quite embarrassed and waved Mary Jo nicely back into her chair. He was glad that only Mary Jo could possibly envision the image that popped into his mind of their molecules wrapping into each other.

Walter then described the methods of paranormal research. There have been and will continue to be thousands of paranormal experiences. The main research method involved observation, and reporting the experience as a case study. Over the past fifty years there had been hundreds of laboratories around the world that had attempted to run controlled experiments documenting and attempting to measure some parameters of ESP. Most of these had inconclusive results. ~~At this time,~~ In Walter's opinion, mankind didn't know enough about ESP to be able to create the proper types of experiments. We had to accept that we were still in an early stage of this field and, just like the early astronomers; we must focus on accurate documentation of our observations. It took centuries of observations before Ptolemy devised his model of planetary motion, and then another 1,500 years before Copernicus simplified it. The members of the Archaeus Project looked at Otto remembering he had made a similar comparison of the study of ESP to early astronomy. Wendell made a conscious note to investigate this further himself.

The best paranormal researchers focused on documenting the experiences that occurred independent of biases toward the people, the culture, or the place in the world. Walter then outlined the important criteria for a case. First authenticity, when the investigator was reasonably certain that the details occurred as the person experiencing the ESP event claimed. And second, and more importantly, the case must be evidential, that is the researcher must be certain that the events described were paranormal, something that could not be explained through normal sensory perception and memory recall.

Walter then related one of the best documented ESP cases. It involved a family in Wisconsin during the fall of 1968. Mrs. Acker Hurth had sent her 5-year-old daughter, Joicey, to the local theater where she was to meet her father and brother to attend a Walt Disney movie matinee. Joicey had missed leaving with her father because she was napping. The theater was ten blocks away, and Mrs. Hurth said Joicey had time to catch her father at the theater before the movie started. Mrs. Hurth wrote later that she was daydreaming while washing the dishes and suddenly she had a terrible feeling of fear. She was startled from her daydreaming just as a dish dropped and broke. She then turned her eyes toward heaven and prayed aloud, 'Oh God, don't let her get killed!' For some unexplained reason Mrs. Hurth knew Joicey had been hit by a car. Mrs. Hurth was so convinced by her intuition that she phoned the theater and immediately asked if her girl was hurt. The theater manager was startled to hear this direct question and confirmed that a car had indeed hit her daughter, but that she was all right and that her father was with her. Joicey later said that when she was struck she called for her mother's help.

Many in the audience were expecting the story to end unhappily. They were relieved to hear the happy ending. Walter continued with an explanation. He said he wanted to suggest two possible explanations. The site of the accident in front of the theater was too far away for Mrs. Hurth to have heard or felt anything physical from the accident. Thus, with the timing of Joicey calling for her mother and the mother reacting to it, one could surmise there must be a new energy form that carried this message. Many people believed this was the main model for most ESP experiences. And some paranormal advocates even supported the notion that this energy form was possible for communicating with the dead. Walter and Mary Jo supported another possible explanation, Mrs. Hurth's unconscious mind created a spontaneous image based on a set of logical facts. Joicey was in a great hurry to see the movie. She was running fast and knew exactly where she was headed with a single focus on her mind. The movie theater was on the other side of a busy street. It was midday and there would be a lot of traffic. In this explanation the ESP experience was not based on a new energy form but on a brain that unconsciously interpreted probabilistic data and created an image using the form of our senses. Otto and Earl looked at each other at the same moment. They could tell that each was thinking the same thing, and that this explanation was a fit to their ESP brain model.

Suddenly Mary Jo startled everyone in the room as she jerked out of a half sleeping state with a squeal. Most had noticed her dozing and wrote it off as monotony with her husband's presentation. Walter looked at Mary Jo and realized that something new had just happened. He helped Mary Jo join him at the podium. Karen went running to the podium as well. She put one arm around Mary Jo and could see she was trembling. Walter had a sense of fear in his face. He grabbed Mary Jo by both shoulders and pulled her toward him. He quietly asked her what was going on in her head?

Mary Jo looked up at Walter. Everyone in the room was frozen. It seemed even the wind outside had stopped to hear what was about to be said. Mary Jo slowly collected herself and looked deep into Walter's eyes. She said she had just seen some things that frightened her. She didn't understand them. Walter encouraged her to talk about them immediately. Mary Jo stammered a few sentences.

I saw a toddler being carried by three men in light brown shirts.

The toddler was old enough to be speaking but only made baby gurgling
      sounds.

I saw an airplane and an ocean.

I saw three men hiding some strange looking money in a garage.

I saw a woman weeping uncontrollably. A man was comforting her.

I saw that man looking at three written notes and hiding one of them.

I saw the man lecturing a large group of policemen.

I saw a toddler's small hand sticking out of the ground in a forest.

The hand was not moving.

And then I jerked and the images disappeared.

By this time Earl had also approached the podium. Everyone was silent and in shock. Images of a kidnapping were creeping into their minds. Karen said what everyone was thinking. She asked Mary Jo if the toddler was dead. Mary Jo just stared back at Karen. Earl then slowly got them all to sit in the open chairs in the front row. Earl stood in front of Mary Jo and said he thought she had seen images from a famous incident in history. At the time, it was called the Crime of the Century. Earl said she had described the Charles Lindbergh baby kidnapping that occurred in 1932, about fifty years ago. He knew many details about the kidnapping because this mansion was being built at that time. The owner, Goodfellow, was especially concerned with the kidnapping and became obsessed with it. Earl said he had always felt that Goodfellow somehow knew more about the kidnapping and murder than what was written in the press. It so affected Goodfellow that he changed his plans for the mansion and added additional features to make the place a safe haven for children. Earl thought Mary Jo saw some of the same images that Goodfellow may have seen so long ago.

Walter looked at Mary Jo. They, of course, remembered the Crime of the Century. Both were still in school at the time. The events of this kidnapping and eventual electrocution of the convicted kidnapper were dates that anyone alive at that time would forever remember. The kidnapping of the 20-month-old Charles Lindberg Jr. occurred from his nursery room the night of March 1, 1932. Over the next eight weeks more than a dozen ransom notes were received. The ransom notes were a source of confusion and controversy amongst the many FBI agents and police force investigators. Handwriting experts pointed toward someone of a German heritage, but others felt there were multiple authors of the notes. A ransom was paid with special issued bills but this somehow generated no new leads. Ten weeks after the kidnapping, poor Charles Jr.'s body was found by chance buried in a nearby forest. Four years later a suspect was arrested, quickly convicted, and sentenced to death. He maintained his innocence and his request for a re-trial was denied. He was electrocuted on April 1, 1936.

Mary Jo sat back into her chair. Now everyone attending this Archaeus Project meeting surrounded her. They formed a circle four deep. She appeared out of breath. She slowly started to talk again. She said she felt that the Crime of the Century was really two crimes. First, the kidnapping and murder of an innocent child taken from his nursery at a moment he should have felt the most secure. And second, the electrocution of an innocent man. She couldn't explain these feelings. She really had no reason to have them. She stopped talking and wanted to find a quiet place to rest. Walter helped her stand and escorted her through the door they had used to enter the Great Hall.

Karen suggested to everyone that the meeting be adjourned. Earl spoke with Otto and everyone tried as best they could to overhear the conversation. Earl said that maybe they did get the wrong man. The public was so eager to have some news. The police and FBI were under great pressure to solve this crime. Because of the poor cooperation between the many investigating departments, there was enough leaked information that the real killers could have found a scapegoat for their crime. Once they found a German immigrant who was a carpenter, basically someone who had a tall enough ladder to reach the second story of the Lindberg mansion, they just had to plant some of the ransom money so he would be caught with it. Everyone, the criminals, the police, the FBI, and even the Lindbergh family, was so anxious to have this case solved that the suspect had no chance.

Karen moved closer to Earl. She added that her mother had heard that a secret Jewish Group, who wanted to turn Charles Lindbergh against the Nazi regime in Germany, planned the kidnapping. Before the kidnapping, little was known about Lindbergh's feelings toward Germany, only that he was proud of his Free Mason membership. Since the suspect was German people expected Lindbergh to exhibit an extremely negative view of Germany, but the kidnapping produced the opposite effect. After the kidnapping Lindbergh leaned toward Germany, openly campaigning that the US should not get involved with the brewing war in Europe. This started a rumor that he was a Nazi sympathizer. Karen commented that these conspiracy theories, as they were often called, in the end held some basis of truth. She often likened them to ESP where the sleeping brain collected facts and assembled them in logical ways to create solutions to problems. At this point Karen encouraged everyone to leave and hoped they could experience the power of spontaneous images themselves someday, just as Mary Jo had unwillingly done that night.

Otto and Earl walked toward one of the side doors to exit the Great Hall. Otto was quietly mumbling to himself but loud enough for Earl to hear it. He wondered how ESP and conspiracy theories might be related. Conspiracy theories evolved to fill in where facts were often missing or hidden. As they were learning from these experts, ESP was closely tied to the hidden, the unconscious part of the brain. Conspiracy theories and ESP encounters came to life from people's thoughts. Both seemed to happen when people needed to get through very difficult times. Earl then added that kidnappings had to be one of the most heart wrenching crimes. Earl and Otto walked through the doorway in silence. It was as if they had some sensation that Minnesota would soon have its own kidnapping that would result in years of an unsolved crime and a nation's grieving.

Chapter 7 -- Mind Over Matter

Great Hall
West Winds Mansion
4 May 1983

Wendell Patel arrived at the Library an hour before the scheduled meeting was planned to begin. He was nervous. He knew his invited guest speaker would be arriving late. He had prepared a long introduction for the meeting to bridge to the unknown time his guest would appear. He chose to build his introduction around his hobby and passion, yoga. He had brought a mat and laid it on the floor in front of the fireplace. He performed various yoga poses and continued them as people filled the room. He had invited many of his friends of Indian heritage, and some of them asked if they could join in exercising. He politely asked them to instead take a seat as his warm-up was part of the presentation. Earl and Otto came separately just a few minutes before the meeting time. They were taken aback to see Wendell in a downward facing dog pose with his backside facing the audience as they took their usual seats. It had been a long time since either of them had been on the floor doing exercises, or just on the floor in general.

Wendell's actions had an interesting effect on the crowd. They remained quiet, but did not seem to be watching Wendell. His poses were embarrassing to watch. As a result, the audience was unconsciously entering a meditation state. Once Wendell felt he had an audience that was comfortable and benevolent, he jumped to his feet. He said that for those that didn't know, he was practicing various movements of yoga. In his opinion, the devotees to yoga had been following the psychic path long before anyone else ever studied it. He said he believed that much of our understanding of psychic science comes from the Far East, especially India, his homeland. Since the beginning of time scholars had contemplated life in a timeless fashion, including all manifestations of the mysterious within the realm of their natural philosophies. Yoga became an outgrowth of these studies. Its name implied a union between man and the higher planes. Its foundation strove for a concentration that produced an unbroken and constant connection between the mind and body.

There were many levels of mind and body control to attain in yoga. These levels were thought to prepare for the physic abilities of clairvoyance, telepathy and envisioning the past and future. What was not as well known was that a yogi could continue to develop psychic powers that went far beyond the physical world. Wendell said it was an amazing list of mind over body situations, and listed them slowly word-by-word.

1. *Anima, the ability to shrink to the size of an atom.*
2. *Mahima, to increase in size as greatly desired.*
3. *Garima, to become extremely heavy.*
4. *Laghima, to become light and float in air.*
5. *Prapti, the ability to bring anything within reach.*
6. *Prakamya, the immediate realization of desire.*
7. *Isitva, the creation of matter through power of thought*
8. *Vasitva, the domination over all objects, animate and inanimate.*

Wendell admitted that these powers at first fell somewhere between the realm of illusion and delusion. But tonight, he hoped everyone in attendance would have a small experience related to the power of Vasitva, the ability to control inanimate objects. It would not be done with a yogi but achieved through an ESP encounter. Wendell went on to explain why he liked yoga as a lead in to ESP. Most of the ESP experiences, certainly the ones that had been illustrated during the past three meetings, were associated with the mind in a meditative or dream-like state. Basically, the beta waves had shut down and the brain activity was dominated by the alpha, theta, and delta wave, indicating the brain was in a semi-conscious state. He was interested in how ESP could also be experienced in a conscious state of mind. Yoga focused on the conscious state of mind. Wendell felt that the mind was not only active in defining reality but also could influence reality. He said that tonight everyone present was about to experience the ESP phenomenon of mind over matter.

As if on cue, which was not possible, Wendell's invited guest burst through the double doors of the Great Hall behind the audience. Wendell was as surprised as anyone in the room by this dramatic entrance. Julie Hook let the doors close behind her as she stood at attention. She was wearing a gray trench coat with a broad rimmed, cane woven hat pulled down onto her forehead and tied with ribbons under her chin. She looked prepared for whatever weather a Minnesota May could unleash; snow flurries to bright sunshine. Today, it was overcast and windy. Her hands were in the pockets of her trench coat as if she was grasping precious items in each one. She was in her 50's and had a nearly round face, pinkish in color, with dark hazel eyes that peered over a thin nose and tight-lipped mouth. She was wearing high top laced boots that were easily seen just below the end of her trench coat. She swiveled her head from left to right and then right to left as she summed up the audience. As if she had concluded some analysis, she started to walk up the middle aisle to the front of the room. Wendell led everyone in a round of applause to welcome tonight's speaker, Miss Julie Hook. Julie smiled and nodded as she walked past the rows of clapping people. When she reached the front of the room, Wendell grabbed her right hand with both of his. He then announced Miss Julie Hook a second time and added she was the matriarch of Warm Forming PKMB parties. Wendell took an open chair next to Earl.

Julie quickly defined PKMB as Psychokinetic Metal Bending and claimed she didn't have a 'party' but had 'parties'. Just before she said the word parties, she stopped and acted out a scene like in a mystery radio show where the action stopped before a dramatic climax. She filled the pause with the sound '*dum de dum dum*' as she tapped her fingertips on the podium to the rhythm of the four words. The action certainly grabbed everyone's attention. Many smiled at the little interruption and act. Julie continued with her history in PKMB and the broader field of psychokinesis or PK as it was nicknamed. She said that the demonstration of psychokinesis, or mind interacting with matter, was more effective when creating an emotionally intense situation. This meant that the individual connected his or her mind with the object to be affected and then commanded it to do his or her will. At PK parties as everyone concentrated on the metal to be bent, she would shout bend, and she then shouted *'bend'* so loud and suddenly it made half the audience start. She said this noise translates the thought into the physical mechanism necessary to now move the warmed metal. The intensity of the specific command is important.

In January of 1981, she began experimenting with this idea by conducting, *'dum de dum dum'*, PKMB Parties. Over 60 parties were held involving over 1,500 people of all ages and backgrounds. Close to 90 percent, *'woooww'*, learned to bend metal using PK with a process called warm forming. This term suggests the slight temperature in the metal increased when it was ready to bend. Approximately half the people who had learned to warm-form retained the skill even outside the party atmosphere.

Metallurgical analysis of warm-formed metal had shown that the two important characteristics of metal were a large number of broken, *'crack-crack-crack'*, crystal structures along the metal grain boundaries and a, *'loooww'* thermal conductivity. Another key factor was that the individual must have been consciously willing to warm-form the metal. He must make a mental connection, *'sessss'*, with the object to be bent and deliberately will it to bend.

Metal with low thermal conductivity stayed soft, *'waarmm'*, for only 5 to 15 seconds. The most difficult task was finding the right moment to add the extra force. Many brittle and otherwise physically unbendable objects, such as plastic ware, have been bent at these parties. A few pieces of stainless steel tableware that have been warm-formed have then broken with a loud popping sound. Some objects have been bent while being held in one hand and not touched with the other hand. One individual had recently been able to hold a piece of tableware in two hands and pull it apart.

Some reported other objects in the room bending by themselves, without being touched at all, '*meesterrreee*'.

Julie paused for moment to look around the crowd. She straightened herself and said she was now going past her own observations and was adding her own opinions about the PKMB capability. A peculiar occurrence at her parties was what she called the first-time effect. A person may have gotten dramatic results the first time he or she attempted one of these activities, but failed the next time he or she tried. In her opinion this occurred because, after bending the piece, he or she analyzed what had been done and, failing to understand it, became a little frightened. She felt that the process of being frightened caused the brain to only respond to the body sensations, and as a result isolate it from the unconscious memories. In order to be successful at metal bending a person had to enter a mental state that was connected to energies beyond the five body sensors. She didn't have a model of how this could work in the mind, but it was what she had observed hundreds of times.

Julie then said that this meeting was not going to be a PKMB party. She added her '*dum de dum dum*' sound again. Instead she wanted to talk about the broader field of psychokinesis or PK as it was nicknamed. PK resulted in the movement or influence of physical objects by unknown physical means. There were a number of people who could move various small objects. Each person tended to specialize in one object. As she listed these objects she added a sound effect with each one. Some could make wooden matches, '*zing*', spin. Others could make cigarettes, '*eek*' lift one end off of a table. Some made long hatpins,'*sss*', swivel on their rounded ends. And some could make a compass needle, '*err*', point in a different direction.

These demonstrations took intense mental effort with the palm of a hand used to focus the concentration. PK also seemed the basis for hauntings and poltergeist activities familiar in haunted houses such as the sound of footsteps, *'clump, clump, clump'*, knocking inside closets, *'knock, knock, knock'*, and lights going on and off, *'click, snap, click'*. Julie agreed that this certainly was crazy stuff but hundreds of unexplained cases of what was called poltergeist activities had been documented over many years. Hollywood had exaggerated many of them into popular horror films. The audience was enjoying Julie's presentation. Her little sound effects made the possibility of these PK phenomena more real.

Julie said she next wanted to illustrate some history, *'oohh'*, to show how ESP and PK were intimately connected. ESP offered great insights into the unconscious and conscious mind, and perhaps, into the possessed mind. PK offered a way that the mind could interact with the real world. She held up three photographs of different boys. She explained that each of these boys was recognized at a very early age to have ESP capabilities. Much of their remarkable ESP childhood experiences had been well documented over the past thirty years. As early as the age of three they demonstrated unusual talents in mind reading, telepathy, and clairvoyance.
They were from three separate parts of the world but their paths had multiple similarities that led each to focus their ESP talent into the world of psychokinesis.

The first photograph Julie held up was of Uri Geller, taken in 1956 when he was 10. She drew out the sound of his name as she held the photo high. *'Uuurrriii'*. The picture showed Uri with his eyes closed as he held a round mechanical pocket watch in his left hand. It was reported that this watch was broken and had not run for over a year. Uri, just by holding it, was able to get it running again. *'Tick, tick, tick.'* It was reported that the watch, after being wound daily, continued to run for several years.

Uri was born in Israel. By the age of three he was recognized as having special skills. He would wake from a nap and be able to recite which places his mother had visited while he was asleep. He continued to do this for most of his extended family members with no encouragement. As a child this mind reading came naturally to him, and he accepted it as a normal part of his life. When he was 10 he began to use his special powers on objects. This extended his ESP capabilities to PK phenomenon and his reputation grew.

Not only was Geller himself gifted in PK demonstrations, but also those near him reported a similar experience that eventually became known as the Geller Effect. Julie continued her act to maintain attention to her story and whispered *'someone else is in my mind.'* In the 1960's Mr. Uri Geller's performed demonstrations of his abilities became tremendously popular in Europe and the United States. After these performances strange phenomena had been reported by various individuals who witnessed Mr. Geller's live demonstrations, or saw him on television, or just listened to him on the radio. These reports stated that, on following Uri Geller's instructions, they too were able to bend metal objects simply by stroking them and willing them to bend, without applying any physical force. Other reports referred to windup clocks and watches that were near the person at that moment, and had not been used for many years, which now started to run, *'tock, tock, tock'.* It appeared that Uri Geller expulsed an inherent latent psychokinetic ability in people that either saw him or listened to him. It was this phenomenon that then became known as the Geller Effect. Julie said she dreamed about obtaining this latent psychokinetic ability herself someday.

The second photograph Julie held up was of Mathew Manning, taken in 1971 when he was 15 years old. In a high-pitched voice she repeated his first name, *'Mathuuuuu'.* The picture showed Mathew Manning in his bedroom where he had written and drawn on the walls.

Mathew was born in England. During his childhood Mathew wrote over six hundred signatures on walls, floors, doors, and scrap paper without any reason. His parents and other observers best described it as an automatic writing method. In one particular week in 1971 more than five hundred signatures appeared on the walls of Mathew's bedroom, *'scribble, scribble'*. The signatures apparently represented people from his village (Linton) that had lived from the 14th to the 19thcenturies. Most signatures could not be validated against the actual style of that historic person, but for those that could, they were a near perfect match. Mathew's automatic writing grew to include signatures from artists as diverse as Picasso, Paul Klee, and Henri Matisse despite the fact that Mathew himself had no artistic ability, *'whatsoooever'*. This strange ability came naturally to him and occurred when he was in a daydreaming state of mind.

Three years after this picture was taken Mathew turned his special abilities toward influencing objects, especially electrical equipment. It was quite common for equipment to have failures when Mathew was around and then work perfectly once he had departed. It was general knowledge with many reporters to write notes during an interview with Mathew because any tape recorder would not function if within ten feet of Mathew. His first flight on an airplane had nearly a disastrous outcome. He was invited to Freiburg, Germany by researchers studying the paranormal. On the flight from London the sensor displays and flight controls were acting erroneously, *'zzzzz, sparks'*, until the pilots asked Mathew to take the farthest seat back in the plane. Today Mathew was still active with demonstrations and publishing books on the paranormal. Several paranormal researchers described him as being the world's greatest in psychic abilities.

Julie then held up the third photograph. It was taken in 1975 when Masuaki Kiyota was 14 years old. Holding the picture up, Julie added a sound effect as she said his name, *'keeyootaaaa'*. The picture showed Masuaki holding a box of Polaroid film in his hands while wearing a helmet of wired sensors as he sat in front of a row of electrical instruments.

Masuaki was born in Japan. At a very young age he began describing events in great detail to his family; events that they had experienced without him. As a teenager he used this supernatural ability, which came naturally to him, to produce images on unexposed film using his thoughts. One of his favorite demonstrations was to expose a picture while holding a new Polaroid film packet. He most often thought of large buildings in Tokyo, but the film, when opened without processing, often showed an image of the Statue of Liberty. He had never been to New York and seen the Statue of Liberty. He described this ability as being able to channel and direct energy around him. Like the other two prodigies, Kiyota willingly accepted invitations from around the world to demonstrate his paranormal abilities at workshops and scientific laboratories. He always portrayed a calm, open attitude even though only about one third of the time was the demonstration a complete success.

Julie then listed why these three boys were so important to the study of ESP. First, they were born in three different parts of the world in three different decades. Second, as soon as they learned to talk, they were demonstrating ESP capabilities with the openness and truth of a child. Third, they willingly agreed to be evaluated by various television and scientific crews. None of them had 100% success with the demonstrations but that was normal to them. They accepted this paranormal capability and never felt pressured to make it any more than the natural phenomenon it was. And finally, each progressed from ESP to PK capabilities. All three became equally expert at psychokinesis, metal bending, producing exotic deformations in spoons and forks that couldn't be reproduced with any type of manual or applied tool effort, 'twissst'. Today, the three continued to demonstrate their paranormal capabilities but were not as fascinating as they were when they were young. The honesty and single motivation of their demonstrations, when witnessed firsthand, had opened thousands of people's minds to ESP.

Julie, herself, had witnessed each of these ESP prodigies in action. She never doubted their special skills. They were successful about seventy percent of the time in accomplishing their PK demonstrations, but it was the experience from the entire setting that was the most convincing. Each had an unpretentious attitude and openness. They welcomed any and all requests for demonstrations of their abilities and feats. People brought objects so they might witness these experiences for themselves. Watches that had not run for years began to '*tick,tick,tick*'. Spoons and forks from home or fancy restaurants were bent, '*oooh*' or twisted, '*eeek*'. Images mysteriously appeared on camera film. In most cases, each request or challenge was accepted and accomplished. Julie became a bit emotional. She said to hear about these three was one thing, to see pictures of what they had done another, but to witness them in action was a spiritual experience. Seeing was believing, but witnessing was the real truth. She paused and looked at the ceiling exhaling loudly. She concluded that most ESP experiences were related to the unconscious mind, but PK was more interesting because it seemed to be coming from the conscious mind. However, she added, that for some witnesses it was difficult to differentiate between an act of consciousness or an act of being possessed.

In that moment Julie stiffened and stared at the back wall of the Hall. Her eyes rolled back in their sockets. She began tipping backwards similar to a wood plank falling to the ground. The audience gasped and some in the front screamed. The screams scared everyone, especially Earl and Otto, and caused people to jump to their feet. Earl and Otto turned to the screams behind them and then turned back to watch Julie as she fell straight onto her back, banging loudly onto the carpeted floor. All noise stopped, every person stood frozen. Earl and Otto looked at each other. They were the closest to Julie and both started to move toward her. Slowly Julie began to rise back up the opposite way she had fallen. One third of the crowd watched while the other two thirds screamed and bolted for the door at the back of the room.

Julie rose up behind the podium where she had been before she had said the word '*possessed*', but she seemed taller. Her eyes were pure black and her hair was standing out as if electrified. Her arms hung limp at her side. Wendell, who had first retreated to the back door, stopped and gained a sense of courage, or perhaps, responsibility. He cautiously walked along the inside wall of the Great Hall toward the front of the room. Everyone remaining watched Wendell as he neared the front row of chairs when Julie suddenly twisted her head and gazed directly at him. Wendell stopped in his tracks and stared at Julie. He said her name pensively trying to call her, "Julie". He tried again, "Julie?" Julie did not respond. Wendell announced to everyone remaining that he felt Julie had been hypnotized.

Wendell leaned into Julie and kept whispering for her to wake up. Suddenly Julie responded. She talked in a strange, high-pitched voice. She asked why was everyone here? She said the mansion had been left in her caretaking. She had intended for it to be a safe place for children but all the children seemed to have been driven away. Wendell grabbed Julie's hand as she collapsed and helped her into a chair in front of the podium. Julie's head fell onto her chest and then she slowly started moving her head side to side. Earl signaled the audience to take a seat at the front of the room.

Wendell stayed with Julie. From his yoga experience he was comfortable being around people who were in deep meditation, hypnotized, or even sleeping. Most could not, but he could tell the difference between these trance-like states, particularly if the mind was open during these times to outside influence. With Julie he hadn't had enough time to feel any difference. Julie slowly opened her eyes and looked around wondering exactly where she was. Wendell again called Julie's name. She looked at Wendell and said, "I'm sorry for that interruption. But it was about the only way to clear the room of those more prevalent to the feeling of fear, *zoooommm*."

Julie then asked everyone to take a front seat on the left side of the room. She explained she had been perfecting a demonstration that was described as '*theee world's greatest illuuuusion*'. They maybe knew it as the Indian Rope Trick. She asked the remaining group to sit straight in their chairs and place their hands on the top of their legs with the palms facing up. She watched them calmly as they all settled into this pose. She then walked to the door behind her, stepped into the small meeting room, and carried out a large basket. As she walked toward the podium she started her sound effects in a chanting format, 'slowly, hearing, slowly, feeling, slowly seeing, slowly smelling, slowly tasting'. She set the basket beside the podium and continued the chanting. In her still slow chant she asked Wendell to stand beside the basket. He slowly walked and stood on the other side of the basket from the podium. She asked Wendell to lead the group with a yoga chant. Wendell asked the group to slowly close their eyes and he started them with a long '*uuuummm*'. He said, "*breath in*" and then started the '*uuuummmm*' chant as he exhaled. By the third time the group was in sync with the breathing and soft chanting.

Julie joined in the chanting and watched the group settle into a meditative state. She then softly said that she was pulling a rope out of the basket and it was extending straight and tight to the ceiling. She asked them, using her sound effect voice, if they could see the rope and all nodded slowly. Then she said that Wendell would start to climb the rope. She described as Wendell grabbed the rope above his head and pulled himself up enough to have his ankles grab the rope tight. He paused and then reached with his right hand about six inches above his left hand while using his legs to push himself upward. He continued with his left hand above his right hand. She asked Wendell to pause and led everyone to inhale and exhale in unison. She then asked Wendell to continue the climb just as he did before. Reach with the right hand, pause, and then push with the ankles till the left hand grabbed the rope above the right hand. She continued the sequence of synchronized breathing and then a rope climb three more cycles. Then she said that Wendell's feet were now above the podium and he should let himself back to the floor. She chanted '*slowly, slowly, slowly*' as if she was guiding him carefully back down the rope. Julie paused for a minute and the room remained silent. She then asked them to open their eyes as the demonstration and lecture were now over. The group slowly began to stir and look around at each other. A few started to clap their hands and soon all were completely engaged in thundering applause, all except Otto. Julie nodded with a thankful smile and Wendell escorted her to the door behind them.

As the group started to talk about the magnificent display of Wendell climbing the rope they had just witnessed, Otto grabbed Earl and said he had seen something different. He said he had opened his eyes when Wendell was supposedly at the top of his climb and saw him still standing next to the basket. Instead of feeling tricked, Otto and Earl were excited about the ability of the mind to create such realistic illusions. This demonstration of mind over matter was certainly a wonderful and tremendous area for further study.

Chapter 8 -- Healing

Great Hall
West Winds Mansion
1 June 1983

Dennis Shilling had been organizing Earl's growing collection of health science related books and medical instruments for over three years. Hundreds of ancient books filled the bookshelves in all the small rooms, and numerous medical instruments were being brought for display in the mansion since the collection had outgrown two sites in the Medtronic office buildings. Dennis developed an extreme passion for the care of these old treasures. He felt that the books were not faring well from the changing temperatures and high humidity due to the changing seasons and the mansion's old design construction. He had encouraged Earl to build a vault under the mansion where a constant environment could be maintained to preserve the books. He enjoyed locating the perfect place to display medical instruments. He often tried to get them operational to experiment on himself. He had some health dysfunctions that he had kept to himself for years and thought maybe they could provide some improvement. Dennis had always felt uneasy around the opposite sex and sensed he was getting worse during the past year. He didn't like this feeling but seemed to have no control over its domination. He knew it was not healthy.

Dennis continued to think about the hidden secrets of these instruments. For the most part when he used them on himself he got little effect other than a warming of the body part attached to the ancient electromagnetic contraptions. He considered finding someone who had claimed to have been abducted by aliens and test them in these machines to see if they reacted differently. Finding the right person, a singularity as Otto would say, for these experiments was more difficult than he anticipated. Now he hoped to find someone that would offer insights into the various archived medical instruments and determine if any could stimulate ESP activity. Again no one seemed to have the passion to study and develop a personal knowledge on these types of machines. Dennis appreciated the ingenuity of the various machines, but he also recognized that baseline medicine steered away from these types of therapeutic devices. Very few medical historians took interest in researching what was considered today as a fringe science and witch doctor medicine. Eventually he sought someone who connected the paranormal with healing but didn't expect that the healing would center on his beloved instruments. In time he found Katie Chittum, who lived in southern California, and invited her to speak at this month's Archaeus Project meeting.

Dennis had spent over two months connecting with Katie. He first sent letters to her and she immediately responded. He experienced no fear during this exchange of letters. Katie was very interested in this secret investigation and one day obtained the phone number for the Bakken Library and made a call to Dennis. Dennis was a bit shocked throughout the conversation and hurried through it as best he could. To shorten the call, he agreed that Katie should visit Minneapolis and would have someone help her arrange the trip. When he had hung up he felt exhausted but had not experienced as much fear as he expected. Maybe her healing abilities were something that could help him.

Yesterday Dennis had met Katie at the airport, actually meeting her right at the gate as she came off the plane, and took her to a hotel in downtown Minneapolis. She explored downtown and City Hall until it was time for Dennis to pick her up. He came about two hours early and decided to stop on Lake Street for a light dinner before heading to the meeting. Dennis was uncomfortable being alone with Katie. It had been a long time since he had been alone with a woman. He wasn't sure how to act and Katie, noticing this, assured him that he was being very gentlemanly and should relax and enjoy this friendly encounter. This made him feel a bit more at ease, but not much.

Dennis showed Katie around the mansion and let her explore the Great Hall where the meeting would be held. Dennis said she could take a seat or prepare in the study that was behind the large fireplace. She said she preferred to sit in the garden outside the main entrance to watch as people arrived. This was uncomfortable for Dennis, but he felt she shouldn't be alone so he joined her on the garden bench. Katie was a very attractive middle-aged woman. While she seemed perfectly comfortable just observing without conversation, Dennis was not and tried several times to start a conversation that was often left hanging. He regretted arriving so early. As people came, they nodded briefly at Dennis as they entered the mansion. Only Earl and Otto stopped briefly to say hello, and Dennis introduced them to Katie. They liked the idea of Dennis having a female friend and didn't want to thwart his chances. Just before 7pm Dennis told Katie it was time to head in. He rushed ahead of her to open the large door leading into the mansion, as well as the side door into the Great Hall.

The room became quiet as they entered. Katie took her place behind the podium as Dennis quickly introduced Katie Chittum from Laguna Beach, California. She had been invited to Archaeus Project to share her experience with ESP, and what she called "healing". Katie thanked Dennis for such a special invitation and waited for him to settle into his seat in the first row. The room was nearly full with almost 75 in attendance. The word about the Archaeus Project was spreading and creating more interest than the group of six had expected.

Katie described herself as in her early 50's and had held various jobs in retail sales and office work. About twenty years ago she became interested in the mental stress many people around her experienced. She said she had witnessed so much anxiety in others that she wanted to help them. As she learned about her own inner problems she realized she could apply this toward helping others. She organized her thoughts to highlight three types of health: physical health, mental health, and psychic health. She felt the health care system in the United Stated primarily focused on physical health with most new drugs addressing mental health. The system did nothing for psychic health, not even recognizing it. She felt that poor psychic health manifested itself into the majority of physical and mental health symptoms. She wanted to treat the cause of health problems, not just the symptoms.

She studied many of the topics of the psychic sciences. There was already a community in California, where she lived, that invested in these studies and sponsored workshops and conferences featuring the most experienced paranormal experts. Each encounter was almost like a fourth-dimension experience where she sometimes felt elevated into another world. She developed a passion for the paranormal and read hundreds of books and magazines.

One of the first topics Katie took to heart was Biorhythms. Biorhythms attempted to predict various aspects of a person's life through simple mathematical cycles. The theory was developed by Wilhelm Fliess in the late 19th century, and had become very popular the past ten years in Southern California. The theory proposed three sinusoidal models: a 23-day physical cycle, a 28-day emotional cycle, and a 33-day intellectual cycle. These cycles began at birth and continued as a sine wave oscillation throughout life. The theory was based on the idea that the biofeedback chemical and hormonal secretion functions within the body had a sinusoidal effect on behavior over time. She had had the opportunity to study the biorhythm model with hundreds of volunteers. She asked each on a given day to score from 1 to 10 their impression of how they felt about nine attributes. She plotted each person's physical cycle with the subjective scores for coordination, strength, and well-being. She mapped each person's emotional cycle from their score for creativity, mood, and awareness and overlaid each person's intellectual cycle from scores for alertness, memory recall, and communication. She chose these nine personal attributes because she felt that these could be indicators or even produce symptoms of a person's stress and anxiety. She believed biorhythms could offer a simple model to help the problems that were making everyday life so difficult for so many. But when she plotted the data it was very noisy. It didn't seem that this model could fit a group of people. She concluded that she needed to find a method that was more interactive with the individual.

Katie paused for a moment and surveyed the audience. She felt they were following her reasoning and ready to move onto the next idea. She said her next passion became Tarot Cards. No one knows the exact origin of the word Tarot but Tarot Cards had been around for hundreds of years. The first decks were hand painted with elaborate figures and divided into four suits. From these detailed decks the German deck of 32 cards and the French deck of 52 cards evolved. Readers of Tarot preferred versions of the early card decks with their detailed imagery and symbols of human desire. Tarot became a visual and poetic language that generated a response to an embodied situation. Each card illustrated what its character was doing and the reader could match it to the subject's situation. It helped clarify what was known and unknown. Being able to articulate between what was known and unknown about a situation was the first step to resolving that situation. Otto shot his hand into the air. Katie asked Otto what he wanted to share. He jumped up saying that he liked the concept of distinguishing between what was known and unknown. Clarifying this uncertainty was the basis for any successful research. Katie felt Otto could continue this dialogue for some time and quickly thanked him for this observation. Otto smiled and sat down into his seat.

She said that Otto's point was very important. After studying and exploring Tarot as a way of healing, she felt it did have limitations. Tarot took on a theme of magic instead of a defined process. It preferred to simply identify the uncertainty related to the subject's concern, rather than address what was causing it. Tarot was very successful for bringing out thoughts, concerns, and events that were hidden in the mind. The images of the Tarot generated an idea from which a self-image evolved. For some people just being able to verbalize and discuss these hidden images opened a path to healing. But for Katie the Tarot was more useful for fortune telling instead of healing. The diversity in the cards and the 'luck of the draw' led the reader and subject to create events and then forecast something about those events. These events were extrapolations from the present from which a decision in the present would lead to. The reader helped the subject clarify these decisions and play out the consequences. In this way most Tarot readers used the cards for helping people find answers to personal questions and impending life decisions. In some ways this could be thought of as a form of healing, but it didn't address the deep facets of psychic health.

The capability of Tarot to illustrate and verbalize hidden thoughts and concerns made Katie think about ways to bring the unconscious to the conscious. This led her to study and work with sensory perceptions as mechanisms to trigger thoughts and events in the mind. She primarily focused on smell. She described working with hundreds of subjects and found that smell was the most successful at stimulating or triggering hidden thoughts. Katie also believed that of the five senses, smell was the first sensory system to be active after birth and the last to fail before death. Smell and breathing were intimately linked. It was also a sense that had an almost immediate memory response. She was surprised how quickly the people in her studies distinguished between smells that were familiar and unfamiliar, and put them into categories of pleasant and unpleasant. Otto was still on the edge of his seat as Katie was speaking. He wanted to add to this discussion and again waved his hand high into the air. Katie stopped and again asked Otto what he would like to add. Otto wondered if the audience knew how fascinating the human nose really was. He asked if they knew that any given smell might contain thousands of different odor molecules? Most of the audience looked at Otto and slowly shook their heads sideways to his question. Seeing their response, he then asked if they knew how many odor molecules made up the scent of a rose? Katie answered saying the complex scent of a rose came from over 1,200 different molecules. As she answered she nodded Otto back into his seat.

The use of smell for healing made sense to Katie. She had collected hundreds of oils that had been extracted from plants. Stories of plants being used by humans for healing were confirmed in recent archeology discoveries. These old stories, handed down generation to generation, described combinations of oil-based fragrances known to produce affects such as calming, awakening, reducing headaches, relieving joint pain, relaxing nervousness, and quieting stomach aches. Katie knew of over 120 essential oils that formed the basis for aromatherapy. But she wanted to extend the use of smell from healing of the physical body to a healing for the mind and soul.

To get to the soul she felt she had to get to the unconscious mind. Since smell is thought to be the last sense to offer some connectivity to the conscious world before death, it also may be a powerful link to the unconscious mind. Through her own studies she also believed smell had direct connection to memory recall. It seemed that nearly all the subjects had certain smells that immediately created a picture in their mind of a previous experience. And this memory function seemed well entrenched into the brain because it was readily evident across all ages, male or female.

She asked Dennis if he might help her with a demonstration. Dennis was surprised by this request but slowly stood, dragged his chair next to the podium, and turned it to face the audience. Katie tried to comfort him by placing a hand on his shoulder and asked him to sit in the chair. Dennis had a mop of long brown hair and wore large plastic framed glasses in a way that seemed to provide a shield from the world around him. He usually had a scruffy beard but because it was summer he turned it into more of an unshaven look. He was comfortable being in front of people when he could prepare and manage the agenda. The situation he was in with Katie was far from that.

Katie reached into her purse, which she had set down inside the podium. She pulled out a cloth bag that was tied with a drawstring. The bag was made of paisley cloth and the drawstring, a bright yellow braid. She opened the bag and took out five small bottles and placed them on top of the podium for everyone to see. She said to Dennis that in her research she found that not everyone had the ability to accept silence. The acceptance of silence was the gateway for the use of fragrances to encourage the unconscious.

Katie stood beside Dennis and held the five small bottles in her left hand and kept her right hand firmly on Dennis's shoulder. It was a gesture to either calm Dennis or-to stop him from bolting out of the chair. Only Dennis knew which was the case. She apologized to Dennis that this exhibition was going to be faster and maybe more aggressive than she normally did, but felt it was needed to illustrate these methods and their connection to the brain. Dennis had shared with Katie earlier, his drug experimentation on himself in the 1960's. She had been quite interested. Now he knew why. Katie said that there were three basic ways to use essential oils: inhalation, topical application, and internal consumption. She was going to use all three methods to illustrate the potentially different effects. She opened one of the five bottles and tipped it upside down on her right fingertip. She held it for ten seconds and then righted the bottle. She then rubbed that drop of liquid into Dennis's right temple. She lightly swirled her finger covering an area about the size of a quarter. After about thirty seconds she repeated the application onto Dennis's left temple. To do this Katie had to stand beside and bend down toward him. Dennis could not help seeing how her white blouse fell away from her body revealing two pushed up breasts held underneath by a white bra. He purposely tried to turn his head but Katie held him in place with her fingers softly caressing his temple.

She could feel Dennis slipping into a calm state. A light head was not the only sensation Dennis felt, he was also having an erection. He moved his hands to hide it as quickly and thoroughly as possible. Katie continued to stoop in front of him as the oils sunk in. Dennis did something he couldn't have done before and looked Katie directly in the eye. She was beautiful, with shoulder length black hair that curled toward her neck. Not one hair was out of place. She wore a light red lipstick and had darkened her eyebrows and eyelids in a way that perfectly accented her hazel eyes.

Next Katie said she wanted Dennis to inhale some vapors that would further open his mind. She opened two bottles, each liquid a different shade of green, and started to wave them slowly under Dennis's nose. She asked him to inhale slowly and deeply. Following her instructions, he began moving his cupped hands back and forth over his ever-growing erection. Katie and Dennis continued for several minutes. Dennis's eyelids began to droop but he kept his eyes fixed on Katie, in her entirety. Katie began to explain that people believed there was a clear distinction between conscious and unconscious states of mind. However, altered mind conditions could confuse the line and cause hallucinations. In the unconscious state dreams evolved from memories of places and people from the past or present. As we awakened, our mind switched from pictorial thinking to word-based thinking. Hallucinations occurred when the mind didn't completely switch from the pictorial unconscious state to the word-based conscious state.

During hallucination, the memory areas of the brain overrode the sensory regions of the brain. The memory areas were now in charge and called upon the sensory areas to deliver a hallucination incorporating the needed sensory information of feeling, smell, sight, taste and hearing. Since all the senses, as well as familiar memories, are involved in hallucinations, perception is distorted for a short while. When the brain exhibited a hallucination it often triggered a memory of a past representation and the person awoke confused because there was an added reality to the hallucination.

Otto almost launched himself out of his seat. He said Katie's explanation of a hallucination matched their brain model of how ESP originated from a memory stimulus when the body sensors were in a quiet state. ESP and hallucination seemed to originate in the same way inside the brain. Earl interjected loudly that there was also a big difference. He said ESP was associated with sane people having paranormal experiences while hallucinations were associated with mental illness and insanity. Katie then suggested that ESP could be used as a way to study the brain so that methods for treating mental illness could be developed. In her work she felt she was focused on minor forms of mental illness such as the anxieties, the worries, the negative attitudes that came from everyday life.

Dennis, unaware of the conversation around him, continued breathing deeply and stroking himself until Katie offered him the final oil for internal consumption. He smiled at Katie and nodded slightly. She grabbed the fourth bottle with blue liquid and put a drop onto her middle finger. She looked at Dennis and slowly put the drop onto his lower lip. Dennis licked his lip with his tongue. He seemed to dislike the taste but didn't try to spit it out. Katie stood silently beside Dennis. He appeared dazed and then asleep. His head lolled back with his mouth wide open. His heavy breathing, now mixed with groans, continued while his hands became frenzied. Suddenly his body tensed while Katie braced the chair to keep Dennis positioned in it. He shouted something sounding like a combined WOW and BABY. His hands relaxed and he cupped his groin tightly. Slowly he opened his eyes and settled back into his chair.

Katie said that she believed Dennis had been healed from a problem he had been having for a long time. While they had been together she said Dennis had not directly stated his problem but she diagnosed it easily. He had a feeling that he was impotent and as a result had shunned encounters with women. She stood in front of Dennis and looked him in the eye. She said that this problem should now go away and he could have a much better quality of life.

Earl was in a state of shock and immediately called the meeting to a close. He spread his arms wide and corralled all the guests to leave through the back door of the Great Hall.

Chapter 9 -- Investigation

(six months later)

West Winds Mansion
Minneapolis, Minnesota
11 January 1984

The Archaeus Project had been meeting less regularly the past few months and was limited to the group of six. Tonight, Otto had asked everyone to be prompt and wear comfortable clothes. He had invited one new person to join the meeting. Dennis had unlocked the courtyard door about thirty minutes early and shoveled some of the snow away from the entrance. Everyone followed the footprints through the snow to find this unlocked door. Earl arrived about the same time as John and then about ten minute later Karen and Wendell expanded the footprints in the snow to the courtyard. It had already been dark for about two hours and a serious cold was locking everything in place for the night. Each step in the snow awakened a sound much like shattering glass. Just as Wendell was opening the door to let Karen lead him into the building, that same sound of shattering glass was coming from around the corner. Otto was stomping, mostly hitting the packed footprints, and trailed by a bundled and hooded follower. Wendell held the door for them as they nearly ran into the warm hallway.

Otto tipped off his flannel hat with the foldable earflaps hanging down and opened his lined trench coat starting with the bottom buttons first as he continued to hold his top-loading briefcase with his left hand. His follower flipped her hood back and unzipped her Columbia winter coat. She fluffed her hair with both hands that caused her shoulder length black hair to fall into place with a slight curl at the end. Under her large full-length coat, the turtle neck sweater and dark pants added to her well-kept, fashionable appearance. Even her fur lined boots looked stylish. Otto introduced her to Karen and Wendell as Carla, saying she would be leading the program tonight. Together the four of them walked toward the small meeting room that was west of the Great Hall and met Earl, Dennis, and John. Everyone said hello and took a seat around the oval oak table.

Otto welcomed everyone to the meeting and thanked them for wanting to meet on this very cold night. He said he'd been preparing for this meeting for a few months and had a plan for the evening. Before he introduced Carla, he wanted to demonstrate some devices he had been working on. He reached into his large briefcase and started pulling out small gray metal boxes. Each had a probe coming out the top, a meter on the front, and red and black screw posts on the bottom. Next, he pulled out four battery packs of D-cells each stacked four across and two deep. Each battery pack was about the same size as a sensor box. He set the four sensor boxes side by side on the table with a battery pack next to each one.

As he pulled a coil of wires from his briefcase he said, "I've been thinking about how ESP leads people to believe in mysterious energy forms. I've built four devices that sense different forms of electromagnetic energy. Two are sensitive to electric fields and two to magnetic fields. It took some circuit trickery to make these sensors sensitive to the specific fields and still operate on battery power. The battery pack provides twelve volts so no worry, they won't shock you if you touch the leads."

Earl looked at these mysterious boxes and said, "What are you going to do with them? Are they communicators?"

Otto shook his head and replied, "No. They are passive sensors. Each responds to a specific energy." He pointed at each sensor and said, "This one picks up low frequency electric fields and this one responds to high frequency electric fields. I am most proud of these other two. They are based on the unbalanced magnetometer I developed long ago for the military."

Dennis cut in, "Really, you've not told us much about that. What was it used for?"

Otto sat back quickly and replied, "Umm. I can't talk about that. It's classified."

Everyone could see the seriousness in Otto and waited for him to move on. He took a deep breath and continued, "This one senses low frequency magnetic fields and the other high frequency. All four have sensitivity in the parts per thousand!"

Only Earl knew what this meant, but in the way Otto said it, the others accepted it as important. Earl believed they were about the best portable sensors ever made. Otto then said, "Once I hook wires to the battery pack, each sensor will be active and I will demonstrate its use. But first I want to have Carla explain what she has planned for the investigation."

John repeated excitedly, "Investigation?"

Carla rose from her chair next to Otto's and began, "I specialize in paranormal investigations. Otto has invited me to lead an investigation in this mansion. I understand you've had some strange events happen during some of your meetings."

"Yes, we've had a full set of fishing bobbers fly off shelves for no reason," added John.

"And someone seemed to be able to see a past event," added Karen.

"And they said I did the Indian Rope Climbing trick," whispered Wendell.

Dennis didn't want to say anything, but he felt for the sake of the Archaeus Project he had to contribute, and said, "And I was healed."

Carla looked around the table as they stared at her. She eventually said, "I've been invited to help many people explain what they experienced. The one common trait of all these people is their combined honesty and bewilderment of what happened. They believed they experienced something, something very real, but they had no way to explain how it happened. I'm not a physicist, or even a scientist. My degree is in psychology. I have been, by some fate, drawn to meet people who have had paranormal..."

Otto interrupted, "Or what I would rather call extrasensory perception, ESP for short, encounters."

Carla smiled at Otto and continued, "Thank you Otto. I met Otto when I was a student at the University of Minnesota and Otto and Viola were and still are legends for connecting science and humanity. I've continued to visit him and he told me of some of your meetings here at the mansion. We both decided a few months ago that we should do an investigation of the mansion for paranormal activity."

"So, I decided to build these sensors as part of the visit, and finally we are here tonight," added Otto.

Carla began to explain her views of paranormal activity, which raised many questions. Everyone wanted more clarification, but in the end, it was still not very clear what an investigation uncovered. Carla seemed to be content to live with the unexplained and take each encounter as it presented itself. She said she always suspected electromagnetic energies to be present during these ESP encounters, but didn't have any way to test for them, until now. Otto, creative and scientific, invented these four portable sensors that they would test tonight.

Otto asked Dennis, Karen, John, and Wendell to each carry one. He had them take the sensor box in their right hand and grab a battery pack in their left. As they held these he connected a red wire between the red terminal on the battery pack and the sensor and then a black wire to those colored terminals. As he was doing this he explained why one should connect the positive terminals, noted by the red color, first when connecting power to a device. By connecting the negative terminal last there was less chance for a spark because of the direction of the electron flow in completing the circuit. Sparks could lead to fires if combustible materials, like a shirt cuff that may be wet from some spilled chemicals, were nearby or even explosions if combustible gases were present. Anyway, there were no sparks as Otto connected the four sensors. As each sensor was powered, a bright light came on, and the needle on the meter pegged to the right. The light slowly dimmed as the needle swung back to the left side of the meter. Since paranormal activity would be tracked in the dark, Otto had added a light that changed in brightness as the calibrated needle responded to the field of intensity the device was configured to detect.

Otto asked the four to walk around the room testing various objects, including people. All of them found places where their sensor read some level of energy. Sometimes they were the same places but mostly each sensor had a unique position for detecting a large signal. Otto explained that ferrous objects and static charge on clothing could make magnetic and electric field anomalies. Once everyone became familiar with their sensors, Carla said they would start the investigation. She led them through the side door into the Great Hall. Only light from the open door was in the room making all the woodcarvings and the upper balcony have a shadowy appearance. The group wandered around the room and the sensors responded only occasionally. Carla asked them to form a circle in the middle of the room so she could urge spirits to be recognized.

She said she knew that one of the greatest ghost hunters, Hans Mocker, had been part of the Archaeus Project. She had attended that meeting. She said her methods were different than Hans's in that she nonthreateningly called to the spirits instead of insulting them, as Hans did. Carla said that spirits should be treated kindly, and in her experience honey was better than vinegar in making friends.

As the group stood in a circle all the sensors went quiet. Carla then whispered using a lower pitch voice, "Are there any spirits here? Would any of you like to show your presence?"

The sensors stayed dark.

Carla paused for a minute and then said, "If you are here, please come near one of these sensors. Or if you have other abilities maybe you can turn one of them on."

The sensors continued to stay dark.

She paused for two more minutes and then said softly, "Maybe you're having difficulty in being able to control these sensors. They are new and built by our friend Professor Otto Schmitt. They will not harm you."

They waited quietly several more minutes. The sensors remained dark.

Carla could sense the group was shifting their weight more often, a sign they were getting restless. She said they would now move into the hallway. They followed her through the middle door of the Great Hall into the hallway. An outside light in the courtyard revealed to them a frozen garden, with faint footprints filled with blowing snow. As they again formed a circle, Carla repeated the same set of questions. This time Dennis's sensor flickered light and stayed on for ten seconds. As it dimmed out, Otto said that Dennis held the low frequency magnetic field sensor. Carla repeated her questions again, but no further indication came. She then began a different direction of questioning.

Carla whispered, "Are you here because you're unhappy? Are you here because something bad happened?"

Dennis's sensor lit up again, but brighter and faster.

Then Karen's sensor began to flicker. Otto quickly said that Karen had the low frequency electric field sensor.

Everyone remained quiet waiting for more reactions, but soon the sensor lights went out. Carla waited another minute and then said slowly, "We are only here to observe. We're not here to judge you."

They stood silently for several more minutes. Only the sound of the wind swirling through the courtyard could be heard. Carla suggested in her normal voice, they should head to another room, one deeper into the mansion. Earl said he would lead them into the basement. There was a storage room ~~there~~ that he considered the darkest and most remote part of the mansion. He also wanted to make sure this investigation didn't get near his basement vault and those secrecies. Earl led the group down the steep, narrow stairway on the east end of the hallway. At the bottom of the stairs Earl turned to the left and pushed open a homemade door of wooden planks. It was quite dark and Otto turned on a small flashlight that he always carried. The empty room was about twenty feet square surrounded by stone and brick walls with an open beam ~~as the~~ ceiling and a cold, uneven concrete floor.

The group followed the dim light from Otto's flashlight to form a circle. After standing for a few minutes with no sensor activity, Carla asked the holders of the sensors to slowly explore more of the room. Almost as if trained, Dennis, Karen, John and Wendell each headed toward a different corner in the room swinging their sensors searching for something. Otto pointed his flashlight down on the floor dispersing light for the others. As John walked toward his corner, his sensor began to flicker. About the same time, Wendell's also did. Otto quickly added that John had the high frequency magnetic sensor and Wendell the high frequency electric field version.

Carla, in her low voice said, "Hello. We would like to know more about you. Were you also with us in the hallway a few minutes ago? If yes, and you can, turn on one of the sensors."

Suddenly Dennis's sensor flickered.

Carla reacted, "Thank you for that. I see you understand our sensors. You now know they are only passive and don't radiate any energy."

Slowly the three sensors that were flickering went dark. Carla waited a minute and then continued in her low voice, "Can we help you in any way?"

All the sensors stayed dark.

Carla whispered, "Would you like to tell us why you are here?"

Dennis's sensor flickered and then stayed half lit.

Carla continued, "OK. Was it something bad that happened?"

At that moment both John and Wendell's sensors started to flicker again. Otto walked closer to Karen wondering if she was holding her sensor correctly.

Carla repeated, "Was it something bad that happened to you?"

Dennis's sensor flashed bright and then Karen's even started to flicker. About thirty seconds later John and Wendell's sensors also lit dimly. The lights of varying brightness shining in each corner mesmerized everyone.

Even Carla showed signs of amazement and whispered, "We have so much to learn from you. Is there another way for you to communicate with us?"

All the lights suddenly went out and Otto walked back toward the middle of the room near Earl and Carla. Carla tensed and began to stammer in a very low, gravelly voice, "Ich nicht do it". She began to collapse and both Earl and Otto caught her. They quickly helped her out of the room and up the stairs. With one on each side they helped her to a chair in the original meeting room. The other four quickly followed.

They all sat dumbfounded for several minutes. Carla had her eyes closed and was acting as if waking up from a bad dream. Karen asked, "Did you understand what was said?"

John replied, "It sounded German."

Dennis said, "I think I heard, Ich nicht do it."

"Yes. Yes. That's what I heard," added Wendell.

Karen said, "What could that refer to? And why did Carla say that?"

Earl replied, "I don't think it was Carla exactly. I think it was that spirit talking through Carla. If I hadn't seen it myself I wouldn't have believed it."

They sat quietly and wondered about what had happened. Earl eventually continued, "Maybe it has something to do with the Lindbergh baby kidnapping. It seems that theme continues to show itself in unusual ways. I think we should all think about this for a while. To me it's not just what happened but also the aura that surrounded us as it happened. There was something connecting us as we experienced these unexpected happenings. Let's come back to this at our next meeting."

Otto collected and disconnected each sensor from the battery packs. The others began to leave as Carla seemed herself again. Earl and Otto helped her to Otto's car so he could give her a ride home.

Chapter 10 -- The Sixth Sense

(Six months later)

Great Hall
West Winds Mansion
6 July 1984

Otto asked to conduct a meeting to analyze the findings of the Archaeus Project thus far. As the organizer, he limited attendance to those who had been present for a majority of the meetings. It was a very hot and humid day. Minnesota weather seemed against the Project. Each meeting had occurred on the most extreme weather days. Even the mansion seemed contrary to the meetings. Mysterious sounds, whirling winds, and oppressive gloom filled the hallways. The frigid cold this past winter seemed to blanket the inside of the Great Hall. And today, the July heat was magnified to an almost unbearable sweat lodge feeling. Interior doors left open to let the heat escape into hallways would mysteriously swing closed. As the group assembled they fanned out hoping the empty space around them would absorb some of the excess heat.

Since no one was living in the mansion Earl left one door unlocked ~~just one door~~ for the small group that was invited to join the meeting tonight. It was the main door leading from the garden courtyard into the long hallway, the same door Dennis had unlocked for the investigation. Otto was running late and was trying every door to gain entry. There was still plenty of sunlight outside so Otto followed the perimeter of the mansion looking for an open door. Finally, he found the one open door in the garden and rushed into the Great Hall from the back of the room. He was greeted with a few waves and sighs as he walked to the podium. Even though sunset was over an hour away the room seemed dark. Lights had been left off intentionally to reduce the mounting heat in the room.

Otto took his place at the podium and looked directly at Earl. Earl had worked with Otto for over thirty years now and was patient and forgiving of Otto's tardiness. In Earl's experience the wait had always been worth the time. But now he was ready to start and encouraged Otto to begin. Otto took a minute to look at everyone individually in the room to recognize them and secretly take inventory of their anticipation. There were about thirty people spaced throughout the room. Despite the uncomfortable heat, everyone was anxious to hear Otto's analysis. No one had any precognition of how this meeting would proceed or end. Only Otto had knowledge of how it would proceed, but even he had no idea that this would be the last open meeting of the Archaeus Project.

Otto began by stating the five senses: Smell, Taste, Touch, Hearing, Sight. He said that years of evolution had perfected these sensors in biological beings. Humans had the best sensors for sight and hearing, but the animal world had better sensors for touch, taste, and smell. Also, the animal world had developed sensors for infrared, electric, and magnetic fields that humans did not have. The human eye had a tremendous capacity for detecting as few as ten photons per second. Human hearing could detect a half-angstrom of vibration, but a scorpion could feel an angstrom of motion. A frog could feel a micro-g of acceleration. A moth could smell a single molecule. Many snakes had infrared sensors that were equivalent to our man-made night vision cameras. Some fish were able to detect the electric field emanating from just 100 electrons one meter away. Pigeons detected Earth's magnetic field to one percent accuracies. Otto was pleased to leverage his graduate studies in zoology whenever he could but he knew he had diverted from the main purpose of this meeting. The group, though, seemed quite interested in these findings.

Otto then focused on tonight's topic by listing the research activities of the Archaeus Project. There were some small experiments that individuals had done but the main results were from the five previous meetings. The February meeting featured, what Otto called, a ghost chaser. The March meeting featured a brain wave electromagnetics researcher. The April meeting featured two paranormal researchers who had left academic positions to pursue their investigations full time. The May meeting featured an expert in psychokinesis, or PKMB, as she had liked to call it. The June meeting featured a healer that used the senses of smell, taste, and touch to relieve hidden psychic health problems. Additional topics of paranormal research had been touched on including telepathy, clairaudience, clairvoyance, Psychometry, mediumship, trance, déjà vu, precognition, apparitions, ghosts, hauntings, poltergeists, psychokinesis, psychic photography, astral projection, out-of-body experiences, possession, automatic writing, Tarot, biorhythms, and even witchcraft.

Otto referred to the scientific method again. It was the only way he knew to order facts and bring them into focus toward a conclusion. It was the method he had used his entire career in developing a Magnetic Anomaly Detection sensor during World War II and then building a database on the magnetic and electric field affects in the human body. The scientific method had evolved over centuries of investigations as a way to determine the fundamental truth. It consisted of techniques for investigating phenomena for the purpose of acquiring new knowledge and was a continuous process, which began with observations about the natural world. Eventually the researchers devised questions about things they had seen or heard. From these questions researchers developed hypotheses that could solve these uncertainties. The best hypotheses led to predictions that could be tested. These tests of hypotheses came from carefully controlled and replicated experiments that gathered empirical data. Depending on how well the test results matched the predictions; the original hypothesis would be supported, modified, or even rejected. If a particular hypothesis became well supported, usually by other researchers having different experimental approaches but producing similar results, a general theory or model would be established.

Physics was perhaps, the purest science because it almost exclusively relied on the scientific method. The worldwide community of physicists would share their hypothesis and experiments, which others would build on. Most technologies related to transportation, aerospace vehicles, manufacturing, and construction evolved from the laws of physics. Other sciences had more difficulty adhering to a complete scientific method approach. Sciences such as biology, chemistry, geology, astronomy, and psychology had more complex interactions that made it difficult to create experiments that isolated the unwanted parameters so that only the parameter under study was being influenced in the experiment. These sciences accepted that without the capability to devise better-controlled experiments, the field would rely on its observations. Certainly, early astronomy, and to some extent present day astronomy, had set the rigor for establishing a science based on observations.

Otto continued to explain mathematics took yet a different research approach that was not based on the scientific method. Because it was not bounded by any physical reality, the field had pursued many abstract and complex problems. The field appreciated and embraced methods that could propose a solution to a problem from insights and conjecture, perhaps even intuition. A good example of this was the early 20th Century mathematician Ramanujan who filled notebooks with answers to problems without showing any intermediate steps at how he arrived at that answer. He produced thousands of solutions in the form of detailed equations that are now, over seventy years later, finally being substantiated with computer modeling and analytical proofs. The field of mathematics welcomed this method for research and building its technical base.

Otto paused and looked around the room. He raised his bushy eyebrows in a questioning look and proposed a question. Which science does the study of ESP most resemble? Otto then proceeded to discuss how he would answer this question. After hearing from the best experts in the world available to the Archaeus Project, one could conclude ESP was not in a category that could be treated like physics. There were too many uncontrolled variables, especially the complicated intermixing of time and memory. ESP experiments could never be expected to hold up to experiments based on the scientific method. The field was full of mostly singularities, things that had happened once. But at the same time, it was possible that one event might spur another. It seemed ESP, as a new field, should fit somewhere between physics and mathematics.

Wendell, who aligned and agreed completely with Otto's description, offered his thoughts. He elaborated on early astronomy where Otto had left off on his early discussion about it. For thousands of years observers recorded the objects in the sky and many civilizations documented their movements. The most famous effort to record celestial events was Stonehenge, built 4,000 years ago in England, which used massive stones to align with significant solar and lunar movements throughout the year. 2,000 years after Stonehenge, in another part of the world, the astronomer Ptolemy created a model for the sun, moon, and the planets Mercury, Venus, Mars, Jupiter, and Saturn. It was quite complex because it used a coordinate system with the earth at its center. This model included lengthy equations based on time and location to compute the position of these seven objects in the sky. These sets of equations were extremely accurate and used for centuries to predict the sun, moon and planet locations in the sky. Then 1,500 years after Ptolemy, roughly 500 years ago, another astronomer, Copernicus, thought to simplify the model by making the sun the center of the coordinate system. This simplification opened another great leap in the understanding of the universe. Wendell explained he didn't see the question exactly as Otto had proposed it, which science does ESP parallel? But instead felt the first question to address was where was ESP in its technical understanding. Which stage was it akin to: Stonehenge, Ptolemy, or Copernicus?

Karen now stood up and asked if she could respond. She paused for a second, saw that no one objected and proceeded. She didn't understand what Wendell was trying to do. She was not interested in going back in time five thousand years. Many agreed with Karen and nodded their approval. Otto used the moment and took back control of the meeting.

Otto said he agreed with Karen and no one was expecting to go back thousands of years in history. From their recent investigations there were plenty of observations readily available to study, and no doubt, more would come. What Wendell had asked was, "Is our brain function model a Ptolemy-styled or a Copernicus-styled model." If it was a Ptolemy-style the Project should expect that it would take a painstaking effort to individually fit all the ESP observed phenomenon into it. If it were a Copernicus style model it would be concise and allow many ESP phenomenon to be illustrated by a simple structure.

Otto reiterated his categorization of problems into either puzzles or mysteries. They had been treating the understanding of ESP as if it were a mystery. This had involved bringing in experts and observing what was known about the various aspects of the paranormal. He posed other questions. Did the observations they had studied fit the brain function model he had shown at the beginning of the Project? Otto had presented the model at the second meeting and he held it up again for everyone to see. Could a sixth sense be defined as generated sensory inputs from the unconscious mind? Did some of the observations not fit the model? He asked the host of each meeting to offer some thoughts to these questions.

John stood up first to summarize the session with the ghost hunter. He said he had learned most interactions with ghosts, or spirits in the afterlife, occurred when people were having a near death experience themselves. Often, they were hardly alert at the time. Also, the experience was remembered as if it had occurred through the normal five senses. The description of the encounters produced the same sensory inputs as if the occurrence happened in the conscious world. The images, the sounds, and the feelings were something that could already have been in the person's memories. The clothes the ghost was wearing, even the jewelry was described as if they were nothing unusual. The sounds were usual sounds, footsteps, doors slamming, and windows opening. It was the manner in which these sounds were produced that gave the witness uneasy feelings. The entire experience was remembered as if it had been a scene in a play. John concluded, in his opinion, the ESP sensory signals came into the brain through the same channels as the body sensors, as was illustrated in the model.

As John sat down, everyone looked at Dennis. He reluctantly stood up and said he had not prepared anything for discussion. The group encouraged him to summarize as best he could the brain wave electromagnetics researcher. Dennis recalled Nick's reenactment of a butcher having an out-of-body experience, an OOBE. It exemplified how the unconscious mind created a memory that could not be distinguished from being firsthand at those events. As Otto had pointed out through his succinct questions, it was interesting that the butcher would experience events that were very close to his own preoccupation of food and eating. As he drifted off during his nap, it seems the unconscious mind created and played into his memory a logical story of the two people closest to him, his store clerk and his wife. To Dennis the more amazing part of this story was how the butcher could have devised such specific details. For example, he described how his employee and friend took a hot dog, and exactly how the cake was served in its unconventional way. The unconscious mind doesn't just assemble and playback a logical set of scenes, but also has the ability to create and extrapolate situations from highly probable possibilities.

Dennis paused to consider what he had just said, which seemed new even to himself. He then said Nick was an expert on what are called brain waves, electromagnetic waves that emanated from the brain. The most common were beta waves that ranged in frequency from 14 to 40 hertz and were evident when the brain was conscious. Nick had demonstrated that during ESP activity the beta waves became quiet, and the alpha, theta, and delta waves were very active. Dennis admitted using the alpha wave enhancing device that Nick had left. He noted that he had experimented and gotten some reactions. In his opinion, the unconscious brain seemed to have its own purpose and hidden motives.

Dennis sat down and purposely avoided any discussion about the second meeting he had organized. But Earl asked him to stand again and comment on the meeting with the healer. Dennis said that this presenter described the importance of smell and its strong connection to memory. She had demonstrated its potential for healing on him. He felt she had addressed his physical health problem through a psychic health treatment. Dennis said he had to admit he was a better man now. He realized when he had had the rare opportunity to be alone with a woman he felt he was not good enough. But now he realized it was safe to be who he was. What that could be he didn't know yet, but was hoping to find out soon. Otto jumped up and cut Dennis off from going into too much detail. Otto said that for the sake of time they should move to Karen's report.

Karen was anxious to add to the conversation. She quickly jumped up and summarized the session with the paranormal researchers. The Uphoffs had retired early to focus on documenting the truth behind ESP. They had written two books that detailed a collection of ESP incidents they had captured through interviews. Their important point was that these encounters were unpredictable and spontaneous. In her understanding, the two most common characteristics of ESP encounters were spontaneity and individuality. Each encounter was so personal to the witness. If one were to consider all these encounters as data one would immediately conclude this was a very noisy system. From a medical standpoint there was no conclusion. However, in her opinion, and of significant importance to her treatment approach, the occurrence of spontaneous images was very real and probably a basic function of the unconscious mind.

Wendell rose to his feet next and summarized the visit of the psychokinesis expert. It seemed that the development of PK ability came from first having a strong ESP ability. There was a significant difference between having ESP and performing PK. The ESP experiences fit what had already been discussed in the model. And nearly anyone could have ESP experiences, basically observing the paranormal. However, only a few were proficient at PK or interacting with objects through paranormal means. The three prodigy children's ESP demonstrations were best done when they thought the least about them. Their success came from an unconscious state. As they became adults they were more conscious of the consequences of their efforts and as a result became less effective at it. In his opinion, the model did not seem to illustrate these PK capabilities.

Otto was pleased with the analysis. The group was working well together. Also, the model was holding up to the research. The fact that it did not include PK phenomenon was actually a good thing. It meant that the model was specific to ESP and it now could be trusted to guide further investigations to test out other aspects indicated in the brain function model. Otto paused for a long moment as he looked over the crowd. It was a technique he used often with his students. He thought of it as pushing in the clutch before shifting gears in those old cars.

Otto stood behind the podium and said he had prepared his summary of the research to date. He read from a sheet of paper[2].

> *We are all very familiar with the loud and even destructive "boom" that occurs when an aircraft or missile moves through the air faster than the speed of sound, pushing the air aside and leaving a void as the air rushes from all sides and collides with itself, creating an impact that can be heard for miles.*

---

[2] Published in the Archaeus Project newsletter, Volume 2, Issue 2, page 5-6.

*Why has no one looked for the analogous miniature
"boom" that should occur if telekinetic phenomena do
indeed cause physical objects instantly to appear,
disappear, or move by some finite distance?*

*Does the appearing object move the air out of its way
before taking the place of the air? If so even a pencil- or
spoon-size void would make a loud bang. Does the object
materialize among the air molecules and have to "digest"
them? This would be hard for even the most extraordinary
of solid metallic objects to accomplish and would require
infinite accelerations and associated forces.*

*Short of accepting a hypothesis of aerodynamics (or
hydrodynamics when the object is under water), how can
we intellectually allow psychically induced motions to
occur with a minimal fracturing of the laws of physics?*

*If we consider the perceptive imagery built out of visual or
auditory or other biologically received information as
being projected, so to speak, onto the mind's imaging
screen, then these images become the reality we recognize.*

*Telekinetic and other unfamiliar paranormal experiences
then may exist as reality in the domain of imaged
perception as an alternative to existing in the physical
world; thus they are indistinguishable from other reality.
This raises new problems, but allows us to seek out the
"psychic boom" as one of many examples of first-order
evidence that there is another working model of reality.*

*Let's think of the purpose for the ghost. As noted in
psychology when someone loses their partner to death they
know **whom** they have lost, but not **what** they have lost. It
seems we have rediscovered that human trait for grieving
to be able to see and feel **what** has been lost. The
unconscious memory never forgets and can create at
important times as part of the grieving for that loss in life.*

When we ask why religions have built doctrine around the afterlife maybe it did not come as a new thought but as a play onto our already ESP based human way of grieving.

The audience sat motionless during this reading. They were not readily following what Otto was talking about. He seemed to be justifying why PK would be difficult to model since some of the observations seemed to defy the laws of physics. He seemed to imply that ESP was a built-in human trait and not interacting with the physical world. The unconscious memory could create nearly any new thought, and as it was not bounded by the laws of physics, it could create somewhat wildly unbelievable experiences. Still motionless, many were staring at the wall behind Otto and others staring at their feet. Many thought that Otto might not be fully aware of what he was trying to say, but in the end they accepted that most likely he was his usual two steps ahead of them. Sometime in the future perhaps they would understand the meaning of this short lecture.

Wendell then asked Otto about the investigation that they experienced six months ago. Not everyone in the room knew what Wendell was asking about, only the original six members participated. Otto said he didn't want to go into all those details, especially since the majority of attendees were not aware of what had happened. But he did offer some short explanations. First Carla was especially gifted at guiding a group on this type of investigation, using timing and understanding of human emotions to build the belief in what was happening was supernatural. His sensors were sensitive to electromagnetic energy and most likely the high frequency electric and magnetic sensors were detecting the old furnace that had an inefficient motor. Slight motions in Earth's magnetic field could trigger the low frequency magnetic field sensor and the low frequency electric field sensor could be triggered by the presence of large capacitive objects such as a human body. Maybe the most convincing part of the demonstration was Carla's gift at changing her voice and timing her questions when one of the sensor holders would shift the sensor to allow it to register a signal. As

Earl had said at the time, the amazing part of the investigation was the aura we all shared during the experience. Somehow, we were all linked and that linking can be explained in other ways besides exchanging physical energies.

Earl was equally baffled by Otto's sonic boom lecture and short explanation of the investigation. If Otto was throwing a curve ball into the group to keep them on edge and thinking, then Earl had his own curve ball he needed to throw at this group. He stood from his usual chair in the first row and turned to face the audience. He motioned for Otto to stay standing at the podium. Earl's face grew long. It was his normal look and created a sense of fear in the crowd. He was, after all, the owner of the mansion, financier of the meetings, and overall motivational manager of the Archaeus Project.

Earl said that each of them had missed something in his or her analysis, something extremely important. Each meeting had had an unexplained occurrence, best described as a mansion manifestation. Each meeting provoked something that had lain quiet for fifty years. His sixth sense was telling him to tread carefully. First, a presence could be felt when the loyalty of the long-deceased housekeeper came under attack. Next, the entire group was put into a trance during which each person created the end of the meeting in his or her own minds. The next guest experienced a series of images reminding everyone of that horrible kidnapping of the Lindbergh baby. The guest speaker, during the fourth meeting, asked where all the children were in a dream-like state. And to make the events even more mysterious Earl had recently learned that the original housekeeper, Ruby, was from Little Falls and probably a school friend of Charles Lindbergh.

Earl was trembling as he was remembering these things. He had bought this mansion because it was being neglected. He had wanted a place where he could build and share a library related to the early study of medicine. He had wanted a place where citizens would be free to explore non-conventional ideas related to improving public health. He felt the mansion was now trying to tell him something else. He needed to think this over. In the meantime, open meetings of the Archaeus Project would be put on hold.

Earl searched everyone's eyes. He wanted to feel their reaction. He saw something that pleased him. His core group of five was exhibiting a determination. Earl understood it, because he felt the same way. He stepped to the side of the room and proclaimed that this experience, these experiments had opened up a challenge he had not anticipated. He then offered an old story about Einstein. Einstein had published the revolutionary concept called the Special Theory of Relativity in 1905. But it took another ten years before he expanded it to be a more complete understanding of the universe and produced the General Theory of Relativity. He was often asked why it took ten years. His common answer was that it took nine years to get the fixed Cartesian coordinate system from dominating his thinking. The foundation of the General Theory of Relativity was based on an abstract space-time curvature model.

Earl felt that their study of ESP must make a similar jump in thinking. Maybe they had to consider the universe as a great THOUGHT instead of as a great MACHINE. He then abruptly left the Great Hall through the side door that led into the hallway and disappeared into the mansion. The rest of the group lingered for a short time whispering to each other and then slowly left the building.

Chapter 11 -- ELF

(Six months later)

Biophysics Laboratory
University of Minnesota
19 January 1985

Otto continued to think about all the Archaeus Project happenings and decided to create a controlled experiment to validate the functional model of the brain that the project had developed. He also had another purpose. Several stoic people from the US Navy had been visiting the laboratory over the past year. At first, Otto, the Biophysics Laboratory Manager and his wife, Viola, of forty-eight years, were quite welcoming to them because they thought the Navy wanted to fund some of their biophysics research. But after several visits it just seemed they wanted Otto to answer lots of questions, questions about the effects of electromagnetic energy on the human body.

Today he would run the first tests of his newly designed and built, what Viola named, The ESP Magnifier. Viola as usual would operate the equipment and record the data as Otto called out the instructions. This new experiment was built on an old school desk consisting of a chair and a writing table attached to its right side. On that writing table there was an electronic box that had lights, buttons, and dials. The box could give feedback to and record inputs from a person sitting in the chair. Attached to a wooden post mounted behind the chair was a large set of Helmholtz coils. This set consisted of two identical circular electromagnetic coils placed symmetrically along a common axis and separated by a distance equal to the coil radius.

Hermann von Helmholtz developed this set of coils in mid-1800 as part of his pioneering study on human vision and hearing. Otto had always felt a connection with Helmholtz since

Helmholtz was from the same area of Germany as Otto's relatives and Otto had been born less than ten years after Helmholtz's death. Otto clarified the definition of the word 'Psychophysics' coined by Helmholtz to describe the effects on human sensing from physical stimuli. Otto considered Psychophysics as the best description for Viola's and his current research in Biophysics. Just as Helmholtz had focused his last studies on the 'unity of the mind and body', Otto and Viola now followed a similar path. But Otto had access to significantly more sophisticated electronics that could extend the studies deeper into the mind than ever before.

Since the many encounters with paranormal specialists Otto thought of ways to enhance or even encourage extra sensory perception, ESP. This led to the construction of this experimental setup that centered a Helmholtz coil over the head of someone sitting in the chair. An electronics box was placed so the subject's immediate responses could be measured. As Otto was designing this chair experiment, he clarified what hypothesis he should investigate. If ESP was only a consequence of the unconscious mind, it would be nearly impossible to study since he had no way to interact with a person that was unconscious. He wanted to explore if external energies could also trigger an ESP experience, but he had to find a way to do this with a conscious subject. It took over a month of working out the steps of the testing, with Viola deciding what was practical or not. During this detailing of the experimental procedure Viola named the chair the ESP Magnifier.

Today they were ready to test their new experiment. To Otto it was obvious who should be their first subject. It was Dennis Shilling. Dennis was quirky enough that if the experiment changed him in any way it probably wouldn't be noticed. Dennis had been to the Biophysics laboratory numerous times before, and during many of those visits he participated in biophysics experiments. So today, Dennis was not expecting anything different. Dennis arrived at the laboratory just before 10am. He said he parked in the public parking ramp on Washington Street and made the long walk across campus to Otto and Viola's new lab space in the Music Building.

As he entered the door he said, "Hello Viola and Otto. Sorry I'm a bit late. This place is quite a distance from where your old laboratory was."

Otto stood up and said, "We were forced out of our old laboratory, mostly against my will."

Viola added, "Otto, you've never been much of a fighter. Everyone knows you're a lover."

Dennis looked at Otto and then out the window past Viola's desk. He seemed embarrassed that Minnesotans would talk this way. Otto continued, "Because of state budget cuts and donors ready to support the construction of a new engineering building, our laboratory was both in the way and no longer relevant, according to the university's administration."

Viola stood and added, "That laboratory was in an old building built right after World War II. It was always planned to be temporary."

"We performed hundreds of tests in that laboratory and discovered significant findings on the electrical characteristics of the human body," continued Otto.

Viola said, "Otto, it still was temporary."

"But we had used it for nearly forty years."

"What temporary means to a state-owned university is different than what it means to average people like us. Anyway, we've been able to set up most of our experiments in this old Music Building."

Dennis thought he might help the discussion, "The place seems good, except I see all the chalk boards have painted-on music staff."

Otto added, "And for both treble and bass clefs."

Dennis said, "It does take up most of the blackboard."

"Otto is able to write and draw on it, and it's not so disrupting," offered Viola. "Anyway, we can't fix any of that but we can start on our new experiment." She opened the door to the electromagnetic testing room and Dennis and Otto followed.

Otto pointed at the wooden chair with the Helmholtz coil suspended above it and said, "Dennis, this is our new experiment. We want you to be our first test subject."

Dennis slowly replied, "OK. What does it do?"

Otto quickly answered, "It won't do anything to you. Well nothing physical for sure."

Viola cut in, "It generates magnetic fields. We've been testing them on people for many years and haven't seen any statistically repeatable effects."

Dennis asked, "But you have seen effects?"

Otto added, "Just outliers. Singularities. One-of-a-kind responses. Nothing like you." Otto said this knowing Dennis would like to hear it, and it would calm him. But he thought to himself that he didn't say specifically that Dennis was normal.

Viola guided Dennis to take the seat in the chair. He had to lean forward and slide in putting his hand on the attached writing table. Once seated, he straightened his back and noticed his head was perfectly centered between the coils.

Dennis swallowed and asked, "Am I going to feel anything?"

Viola replied, "No. And you can keep on your glasses as they have plastic frames. But if you were wearing earrings I would have you take them off."

"Why did you ask me about earrings?" asked Dennis as he glared back at Viola.

She quickly replied, "Sorry, it was just a bad joke. Please, never mind."

Otto had been checking in the back of the room and walked toward Dennis and said, "I've been planning this experiment for the past six months. As you know we have been investigating ESP for over three years now, but only in a way to observe it."

Dennis replied, "Yes. We've had many interesting meetings."

"Well I think I have a way to encourage it," stated Otto.

"Encourage what," asked Dennis.

"ESP," cut in Viola, "Otto believes he can encourage ESP while the person is conscious."

Otto continued, "Extrasensory perception, as we have concluded with our brain model, is caused by stimulus from the unconscious brain that literally paints a story in our memory by feeding signals to our sensory cortexes. These signals must be based on electric currents since electromagnetism is the physics behind how our body functions. "

Dennis added a thought, "Are you sure? Our brains seem so strange in how they work and fail that I was thinking quantum mechanics had to be involved."

Otto replied, "Quantum mechanics is mostly a mathematical formulation and so far only applied to understand extremely small scale and very cold, near absolute zero, material behavior. It may play a role but we are years away from being able to do quantum mechanical experiments on our brain."

Dennis said, "Once, in January wouldn't you know, I locked myself out of my house and got so cold I was seeing myself getting smaller and smaller."

"Dennis, maybe you should donate your brain to science. It's not wired like most."

"And that's why you are here, Dennis," added Viola trying to get these two to focus on the task at hand. "Well, I don't mean dying and donating your brain, but for science. Otto we're ready to start the experiment."

Otto walked in front of Dennis and said, "Thank you for volunteering to help us."

Dennis cut in, "Am I going to remember any of this?"

"Dennis, we expect you will not feel a thing, but we hope you will have some sensory experiences that you can indicate to us through this response box," added Viola as she pointed to the small plastic box with lights, switches, and knobs on its top. She straightened the cable letting it fall to the floor and followed it to its connections into a rack of electronics about ten feet behind the chair. She walked around the rack of electronics and sat in a chair with her left side facing Dennis.

Otto continued as he stood in front of Dennis, "What we have here is a Helmholtz coil that can produce a very uniform magnetic field across most of your brain. We can generate various strengths and frequencies of magnetic fields."

Dennis responded, "You mean like Earth's magnetic field."

"Well, not exactly. Earth's field is very uniform in intensity of a half a Gauss and relatively stable. We will generate magnetic fields that are much stronger than Earth's field and will vary in time. More like an extremely low frequency radio wave."

"I was with you till you said, extremely. I'm not sure I like that being directed into my brain."

"There's nothing to worry about. It's just an expression for frequencies lower than one hundred hertz. Most radio waves are in the thousands and even millions of hertz. Physicists use extremely low frequency, or ELF for short, to describe the range of waves we are experimenting with today."

"But they're going through my brain."

"Dennis, there are hundreds of radio waves going through your brain, body, this room every second. Electric power companies have been studying this for years, and we have done many studies ourselves."

"And now the US Navy also seems to be interested as well," added Viola from behind Otto.

Otto ignored the comment and continued, "In those studies we were looking for physical health effects. Nothing was ever conclusive. But with this experiment we want to study effects on ESP."

Dennis looked at Otto, "You mean mental health effects!"

"No-no-no. Mental health effects come from what I want to describe as mis-wiring in the brain. We just want to stimulate signals in that wiring, not change it."

"How does that work?"

"As described by Faraday's Law of Induction, a changing magnetic field passing through a material will induce an electrical current. Lenz's Law then defines that the current will have a direction to oppose that change. In essence a changing magnetic field across your brain will produce small electrical currents inside your brain. We hope some of these currents may stimulate the cortex from which you would sense a response."

"You mean something will just pop into my head."

"Well, yes. And by having you nod or raise your chin, or turn your head left and right we can affect different aspects of your brain."

"Will I have visions like I did in the 60's at that commune where I spent one complete summer?"

"We're just trying to do some simple studies; hallucinating drugs is another level of complication."

Viola cut in, "OK. Otto the equipment is warmed up. We are ready to record."

Otto said, "OK Dennis, we need to reduce your sensory inputs as much as possible so we are turning off the lights and we've made our equipment deadly silent."

Viola added, "Otto. It's not deadly. Sorry Dennis."

Otto continued, "It's just an expression. We've made this room so sensory empty that you should not feel, hear, see, smell or taste anything."

Viola said, "And on your control box there are four small lights and five buttons. The lights will direct you to move your head and the buttons are for you to push if you have a sensory response. The buttons are from left to right in alphabetical order of feel, hear, see, smell and taste. Don't worry if you mix them up as you will be in the dark. It is better to push one button than not push the right one."

Otto added, "And the lights direct you to move your head. They are arranged in a diamond and when the top one is lit we would like you to raise your chin as high as possible. And when the bottom one is lit to lower it as far as possible. For the light on the left we would like you to turn your head to the left and vice versa for the light on the right."

Dennis looked at the box, "It seems simple enough."

Otto started to walk to a chair behind Viola and said, "Oh. And when the light goes on we would like you to move your head as fast as possible and hold it there till either the light goes out or another one lights."

Otto had wired the entire room controls into the electronics rack. As Otto sat down Viola said, "Cutting all environmental controls. Going dark. No sounds please."

The room went into complete darkness. The three sat in this sensory dead setting for five minutes. Otto had chosen a Saturday morning and expected the building to be empty of any visitors. In the dark, Viola moved her hand to touch a button. The top light on the control box in front of Dennis just barely lit. Dennis jerked his head up. He didn't sense anything and sat still. The light went out and he relaxed his head to a straight-ahead position. The lights repeated one at a time in a clockwise pattern and he completed the head turns as instructed. He did not sense anything and provided no response. This continued for another ten minutes with a light going on about every thirty seconds and staying on for twenty seconds. During this time magnetic fields were being changed but only Viola knew which frequencies or strengths were being set for each light change. She had worked out a random pattern of tests that she controlled by switching four different knobs between ten different positions. The responses from the five switches on the arm chair control box were input to a rolling five channel chart recorder.

After ten minutes, Otto tapped Viola to speed up the tests. Now the lights didn't go on and then off but another would light right after one went out. It started first having Dennis snap his head left and then five seconds later snap it right. And then right to left and back again. This repeated for two minutes, and at one light change Dennis pushed the button indicating he heard something. The test lights stayed off for a minute to let Dennis rest. Then they began the same jerking motion but for nodding his head. He snapped his head up and then down at five second intervals for two minutes. Dennis kept at it as best he could. Otto was pleased. He figured Dennis would stay at this nearly impossible task as well as anyone. Dennis was known for his determination and often would overrule any logic when being asked to do a task. During this set of tests Dennis hit the hearing and feeling buttons each twice.

After ten minutes of this format, Otto tapped Viola twice. Now the tests speeded up even more. The snapping of the head was in a complete random pattern at the amazing pace of one-second intervals. Dennis's determination took an even stronger hold. He barely had the time to hit buttons but he did. He lost track of how many times and which buttons he hit. He was in an autonomous state focused totally on the lights and his reactions. Eventually he only envisioned his brain moving and sensing deep inside it any stimuli. Finally, Otto yelled, "Cut!" and Viola flipped on the room lights.

Otto rushed to Dennis who seemed extremely dizzy and slowly moved his head in clockwise and counter clockwise circles. Otto said, "Dennis, that was amazing. You performed exceptionally well. We seem to have triggered some ESP effects in your brain."

"Well, I hope that was helpful to science. I didn't even feel any warmness from the intensity of electromagnetic radiation racing at the speed of light through my brain. I now think I need a nap though."

Otto helped Dennis out of the chair, "You've given us a great start. We'll need many more tests subjects before we could hope to see any pattern, but the fact that you gave us a couple hits is very encouraging."

Viola walked to Dennis and took the control box off the writing table and said, "Dennis, thank you again for being such a determined and helpful subject. We would like you to join us for lunch. I think it best you relax for a bit and have something warm to eat."

The three went into the office area where Viola warmed up some soup on a hot plate and served it with sandwiches cut into thirds with a steaming green tea. They talked about the weather and what new plans Earl had laid out for adding to his medical instruments collection. Dennis did not want to talk about that experiment anymore.

(six months later)

Chippewa Tavern
Clam Lake, Wisconsin
28 July 1985

No one living in the area knew when the US Navy turned on its ELF transmitter. But for the past two months many reports of extrasensory perception were being reported. These reports, most often only shared with relative and friends, eventually made their way to Otto and Viola's laboratory. Viola logged the place and time for each report and soon a pattern was emerging. It covered an area about fifteen miles square centered on Clam Lake, Wisconsin.

For almost twenty years the Navy had secretly investigated building an electromagnetic antenna for communicating with submarines in northern Wisconsin. As these paranormal reports started coming in and guessing the US Navy had built a new facility near Clam Lake, Otto started using his old Navy contacts to see if he could learn more. Otto and Viola had worked with the Navy during World War II to help defeat the German U-boats. Viola was against Otto getting involved again with the Navy since the work they did was still classified, and they both remained under secrecy agreements. Otto said he could investigate this without disclosing their past. As he made phone calls to colleagues who had research contracts with the Navy he was able to piece together a plausible reason for the enhanced ESP near Clam Lake.

The Navy had code named the plan, Project Sanguine. Its purpose was to establish a new communication method with nuclear submarines. It was built and apparently operating. As best as Otto could surmise the system consisted of two fourteen-mile long wires orthogonal to each other that formed a large X. A building was at the crossing point and had multiple power lines running to it. The communication method was to generate extremely low frequency radio pulses that could travel through the earth and be detected by submarines deep in the ocean. Today, Sunday afternoon, Otto and Viola had been invited to Clam Lake to meet with a group of concerned citizens. As Otto and Viola drove on this bright sunny morning, Otto kept thinking about that name, Sanguine. It was a seldom-used word meaning bullish.

Otto and Viola pulled into the parking area of the Chippewa Tavern in Clam Lake. They had arrived a bit late and the parking area of this traditional Wisconsin tavern was full. Otto found a spot down Clam Lake Street that was also lined with cars. They hurried into the tavern to find the place full of people. Everyone had a bottle of beer or two as they sat at the bar or at a table. The organizer of the gathering was also the owner of the tavern. Mark Strewlow had learned of Otto and Viola when his niece had studied at the University of Minnesota. She majored in biology and spent many donated hours helping in their laboratory. Mark had called Otto and asked if they might come to Clam Lake to answer some questions that were stewing in their little town. They figured Otto and Viola could complete the trip in one day, needing about four hours each way and about two hours for the meeting.

As Viola and Otto entered the tavern, the conversations stopped, and Mark literally ran to meet them. He said, "Hello. You must be Viola and Otto. I'm Mark Strewlow and we have a large group interested in learning more from you."

Otto replied, "Well, thank you. I'm not sure we have many answers but we're always interested in new situations."

Viola added, "We've always been sympathetic to anyone wanting to tell their story, describe what they feel."

Mark ushered them to the back of the room behind the pool table where they could be seen and heard by everyone in the bar. Otto stood as Viola took a chair and sat beside him. Otto estimated there were about sixty people present. Mark asked Otto to start by just sharing what he knew about this Project Sanguine.

Otto did not talk about the choice of the word Sanguine and especially not about Viola and his experience with the Navy. He launched into a discussion about communications and the probable need to communicate with nuclear submarines. He said that nuclear had two meaning for these submarines. First, they were nuclear powered which gave them the potential to run without refueling for many months, the exact number was classified. As a result, they could stay submerged and hidden, except to communicate. The second meaning was that these subs usually carried ICBM nuclear weapons that could be launched while still underwater. So, in Otto's estimation, the Navy would have an important need to be able to communicate with these submarines while they stayed submerged. Once they surfaced they gave away their location, but underwater they could travel for months and be impossible to find. Seawater was a strong attenuator of radio waves and only low frequency waves had less chance to be detected. In fact, the band of frequencies for this type of communication was called extremely low frequency, ELF, waves. A few people asked some clarifying questions, but for the most part the group was not interested in the physics of what was being done, they wanted to know what it was doing to them, to their chemistry and biology.

Otto then explained that the human brain emitted four ranges of frequencies itself. These were called the beta, alpha, theta, and delta waves. Only the beta waves were emitted when the brain was conscious and the other three patterns were emitted during different stages of being unconscious, or what most referred to as sleep. Not much was known about the purpose of these waves and maybe simply reflected the electrical activity in the brain. Viola could sense that the group was becoming impatient with Otto and seemed disinterested in what he was saying. Viola stood and moved closer to Otto. He stopped and asked if Viola would like to add something.

Viola paused for a moment and then said she had been recording the numerous reports of paranormal activity coming from this area recently. She and Otto could explain what they knew of ESP and what impact it might have, and later they could meet with anyone who wanted to share his or her own experience. Otto gave a short explanation of the Archaeus Project and its efforts to investigate ESP. The Project had invited and observed many different types of paranormal happenings. From his standpoint the major finding so far was that paranormal events were magnificent examples of the capability of the human brain. ESP was a mechanism the unconscious brain used to plant new memories in the conscious brain. They had developed a brain model where the brain itself could generate sensory inputs such that the conscious brain could not distinguish from those that had come directly from the five body sensors. The brain constructed a memory that was as real as if it had been experienced while in a conscious state. They did not have much evidence that external energy could trigger ESP, but they were beginning to study this.

Viola said that over the past six months they had developed and ran tests to try to stimulate ESP in a conscious brain. She said the data was very noisy but a few test subjects had observed extrasensory inputs. They exposed the brain to ELF radio pulses at various angles and sporadically a hearing or feeling sensation was logged. In her opinion it was highly probable that being exposed to the ELF antenna transmission could produce ESP. But because there were so many variables it would be nearly impossible to conclude it absolutely. Viola paused for a moment to collect her thoughts. She then tried to address what was directly on the attendee's minds.

She began, "I don't think being exposed to ELF radio pulses is physically harmful. And we know that having ESP experiences is also not physically harmful. So, if ELF is triggering ESP effects then it most likely will not lead to a physical problem. However, I can't speak for the emotional effects. Generally, ESP was reported as pleasant experiences. Having visions of events happening remotely or discussions with lost loved ones was usually told in a welcoming, encouraging manner."

Otto added, "We don't know exactly the waveforms that the Navy is generating in your area, but I have a pretty good idea. Our work has never uncovered any short-term physical problems from being near these types of electromagnetic fields. However, we have not been testing long enough to know if there are any long-term effects. This would take years of testing with the same subjects. Unfortunately, some of you may be living that long-term study."

This got people murmuring and Viola tried to step in to calm the situation. They had been asked to come and offer some scientific advice, not start a demonstration against the Navy. Demonstrations had already been held several times and were having no effect on changing the situation. She lowered her voice so people had to be quiet down to hear her and said, "We really feel sorry for you. We wish there was more we could do to help you deal with this situation. It isn't fair. You chose to live here being surrounded by tranquility and nature. Now it seems the government has invaded your community, which is just the opposite of what we the people want from our government. I know it feels like we haven't told you much, but unfortunately, it's all we know at this time. Really, we can't say that there should be any physical effects from living near this ELF radio transmitter. And as for your emotional and psychic health, that has had even less study. The best I can tell you is that if the main outcome is more frequent ESP experiences, this should be thought of as a positive step in your psychic health. Now if any of you would like to have private discussions, Otto and I can stay here for another couple hours. Thank you for welcoming us to your community today."

The audience gave a soft round of applause and a few stood to head toward Otto and Viola. Mark came forward and shook both their hands and thanked them for the visit. He motioned them to an empty table and asked if they would like a drink. Both said that a glass of water would be welcome. For the next two hours individuals, and a few groups of two, took turns at Otto and Viola's table. They heard many stories ranging from unusual headaches, to cars not starting, to visitations from dead relatives. Otto and Viola mostly listened and offered a scientific explanation wherever they could. Most people seemed happy to talk with someone about their concerns and experiences. They got a late start back to Minneapolis and decided to spend the night in Hayward, Wisconsin. Viola talked Otto into visiting the Fresh Water Fishing Hall of Fame the next morning before driving back to Minneapolis.

Chapter 12 -- The Library Becomes a Museum

(One year later)

Great Hall
West Winds Mansion
9 July 1986

The Archaeus Project had been operating underground for the past three years.
They now met inside the vault of the West Winds Mansion. In 1981 Earl had the vault built into the basement of the mansion to protect his most valuable books and machines. Whatever else it was protecting only the members of the Archaeus Project knew. The public knew very little about the Archaeus Project activities and was unaware of the subtle experiments that were being performed on unsuspecting participants.

The small group continued to discuss how a better understanding of the mind would lead to developing medical practices that could double the quality of life at half the cost. They were confident that the brain function model they had developed was accurate enough to be used for further investigations. Four of them (John, Karen, Dennis, Wendell) took an active role in bringing new results and ideas to the discussion. Otto and Earl served as a panel of experts to assimilate the new findings into their understanding of ESP and modify the brain function model accordingly.

Dennis kept exploring with the low frequency electromagnetic generator that Nick had left behind. Each month he reported the results of his covertly performed experiments on unsuspecting subjects. He hoped he could stimulate an ESP experience. So far, he felt that at times he assisted people to more easily enter a meditative state. He could only record the effect of this on his subjects. Being, in his own words, a pencil necked geek, the feelings of empathy didn't come naturally to him. Sometimes he pushed experiments too far. One night he was cleaning the hallway in the physics building while there was a Quantum Physics class in session. He set the device to enhance Theta Waves and left it running at the back of the room. When he came back to the classroom everyone was asleep, including the professor. Dennis immediately shut the device off and hid it in his mop pail. Then he slowly went around the room, nudging people till they stirred back to life. He was able to arouse everyone at about the same time so the entire episode went undetected. Another time his device had just the opposite effect. He was dining in Dinkytown, and adjusted his device for the encouragement of Beta Waves. He rested it on the table as he slowly ate his gyro sandwich plate. Again, he said he should have realized the effect earlier, but by the time he noticed his surroundings the place was full of noise and discussion. Customers were complaining about the length of time it took to get served or that the wrong food was delivered. The place looked like an organized riot if there was such a thing. There were no physical interactions, except for a few plates being clanked with silverware, but there certainly was plenty of verbal activity. The wait staff, in a rush to clear Dennis's plate, swept up his black box transmitter without his knowledge. As the restaurant quieted down he heard loud noises coming from the kitchen, pots and pans clanging together and knives slamming cutting boards. He abruptly realized his transmitter was missing and rushed into the kitchen. Dennis grabbed his special device just before it was headed into a dishwasher and slipped out the back door of the restaurant. After telling his story, the Archaeus Project passed a rule that he was not allowed to bring this device into the vault.

John continued to focus on comparing beliefs associated with religion and beliefs associated with paranormal phenomena. Because of his high position in the Catholic Church he often had opportunities to interact with organized groups both inside and outside Minnesota. He quietly kept a log on individual people he met. He was able to chart data through simple observations or with properly phrased questions. He scored each person from 1 to 10 in three categories: conventional religious belief, church attendance, and belief in paranormal phenomena. He found a positive correlation between religious belief and paranormal belief. Basically, if a person was above a 7 in religious belief they were also above a 7 in paranormal belief. However, greater church attendance was negatively correlated with paranormal belief. If a person was above a 7 in church attendance then they were below a 3 in paranormal belief. Otto interpreted these findings and said that if one believed in God then they also believed in ESP, but if one went to church regularly, as part of their belief in God, then beliefs in other non-worldly thoughts including ESP was suppressed. Earl added his own observation. He concluded that religions like to promote miracles, but only their own.

Karen established a program using biofeedback to help children gain control of bodily functions that had been damaged or missing since birth. In controlled studies, her research had documented the abilities of children to control certain physiologic processes voluntarily that were previously believed to be autonomic. As children succeeded in accomplishing the desired control, they often described images that arose spontaneously, which they used to effect the changes. Images varied from child to child. She had convinced her colleagues that understanding the source and nature of these images would lead to treatments better than the machines to which the children were connected. The machines only responded to a stimulus that was triggered by the images. With the foundation of their brain function model Karen hoped to understand how to preserve imagery skills developed in childhood. Adults could then recall these skills for self-control of physiology. Karen was planning a major research program that would include thousands of detailed experiments covering numerous variables such as age, development, genetics, diet, biorhythms, medication, climate, the presence of adult coaches or child therapists, classroom teaching methods, the effects of television, remote-viewing skills, and other unknown variables. Otto suggested that the answer might more easily come floating into someone's consciousness as an unexpected but welcome image.

With Earl's encouragement Wendell focused his investigation into ESP and kidnappings. Their public meetings had reopened a door to the Lindbergh baby kidnapping. The first meeting awakened the ghost of the first caretaker once they began questioning her ability to keep children safe. An unexplained disturbance in the hallway surrounding the north and east walls caused a rapid-fire fall of fishing bobbers. At the third meeting one of the invited experts slipped into a trance and saw images related to the kidnapping. The fourth meeting brought attention to missing children and the fifth meeting examined criminals involved in such inhuman acts. Wendell used his political influence to learn details of the Lindbergh baby kidnapping. Charles Lindbergh was a public figure and offered almost daily press briefings for months after the kidnapping. He took the lead in organizing the investigation because he felt the local police were lacking. Because the crime was labeled as a national disaster, federal organizations became involved and to this day the FBI takes a prominent role in investigating any kidnapping. Wendell felt that in the beginning the investigation was determined to find truth and justice. But when the case became difficult to solve, truth gave way to the pressure to solve the crime. The authorities rushed to sentence and execute the suspect for the sake of closing the case. Once the case was closed, the world could move on. However, today there was still an open discussion about truth and whether the right person had been arrested. Wendell investigated hundreds of kidnapping cases and each had some ESP aspect arising from witnesses or people related to the case. Earl and Otto pondered these facts as Wendell uncovered them. Their first goal was that the Archaeus Project would provide a breakthrough in understanding ESP so that it could improve medical practices. They also hoped that their findings would be useful for criminal investigations, especially kidnappings.

In the end, after more than three years, the Archaeus Project was losing its momentum. The group had no end of strange happenings to investigate but these new results were not leading toward a better understanding of ESP. Their model provided a great framework to illustrate a system-level mechanism based on the unconscious creating thoughts that included the same sensory detail as conscious thoughts. How and when the unconscious was able to generate these events was a mystery. An even greater mystery was how the unconscious could produce such accurate and real encounters for both past and future events.

Otto felt the occurrence of simultaneous thoughts in individuals, especially twins, could be likened to computers. The more the computers were alike, and the more they were fed the same data or used the same memory, i.e. experiences, the more the outputs matched. Computers were deterministic to a fault. For example, as people worked with computers they inevitably got a wrong result. They knew it was wrong because either the computer program was being tested against known problems and the output didn't match what it should or the output wasn't within the bounds of a logical estimation. In these situations, people would try to run the program again but soon realized that with the same inputs the computer program would output the same results. Computers began to be described as 'garbage in, garbage out'. Otto realized his use of the term garbage was what most people thought of ESP, but he felt there were treasures in this garbage yet to be discovered. Could the human brain have a similar complication that not all data being processed led to logical results?

The group purposely avoided any detailed discussions or studies involving secret government organizations, unidentified flying objects (UFOs), and space aliens. They were well aware that many paranormal experts and publications often extended the discussions to include these topics. Stories of UFOs and aliens offered believers embodiments of the ultimate paranormal powers. These stories could illustrate new energy forms, unbounded forces, and undetectable mind-to-mind communications that they believed were the basis for all paranormal phenomena. Otto pointed out many times that even with the extensive array of precise instruments these paranormal energies or forces had never been detected. This was primarily what drove Otto to develop a model in which ESP was possible within the mind itself. Earl realized that this outcome was not what experts in the occult would embrace. The field was too broad, had too much history, and too many stories to be explained in such a pragmatic way. It included millions of documents that now formed a dogma and enlivened the mysteries of the occult. Earl also realized that science, especially medical science, would not readily adopt the findings from the Archaeus Project. Their brain model for ESP was mostly confirmed with isolated events, singularities. For this model to be adopted it would need parallel studies and repeatable results.

This realization of the difficulty to make progress in paranormal research led Earl to think of other, more productive, uses of the West Winds mansion. The research had achieved some success and would serve the next wave of research that followed. Paranormal research will take many steps before reaching a scientific level equal with physics and astronomy. And much of that will need to include others. For now, Earl moved to redirect the mission of the mansion toward a broader public good. He would soon rename the Mansion from The Bakken Library to The Bakken: A Library and Museum of Electricity in Life. The emphasis would shift from being a library to a museum. He felt a museum could appeal to a broader group that would include adults and children. His vision was to see children in a safe place, a welcoming place, and experience science in a hands-on atmosphere. He knew that initially the museum must focus on the normal science and especially the exploding field of medical devices. But he hoped someday it would also encourage the study of the outliers, the singularities that may offer the final frontier for establishing the best possible public health.

For the years that they met, the members of the Archaeus Project felt the discovery of the science underlying ESP would be equivalent to the scientific discovery of atomic energy. They felt a personal obligation to society to manage this tremendous force, before society let the 'genie out of the bottle' and created another world destroying situation. The Archaeus Project was not able to find this science.

Or did they? Do they still meet? Who continues to meet? What have they found? Does the hidden vault still exist? Has Earl Bakken, following the themes of the original mansion design, created hidden passages and enabled the secret studies to continue? Has ESP been embedded into the culture of Minnesota? These are questions that will need their own project to find the answers.

Afterword

The mystery surrounding the "Crime of the Century" as it was called at the time, the Lindbergh Baby kidnapping, makes for a parallelism to the study of the paranormal. Kidnappings of children, especially babies, are the most horrific crimes. There is no greater situation that demands a rapid determination for truth. Solving this crime, which happened in the midst of a well aware and structured society, was very complicated. The crime was first treated like a simple act. As the investigation progressed the case became more perplexing. Not just because of the thousands of leads that led to dead ends, but also because of the way the numerous investigators misrepresented the data between truth and hoax. Eventually everyone, including the Lindbergh family and even President Roosevelt, wanted the case to be simply solved and then go away.

The occult continues to be one of the great mysteries of our generation. Its past is full of both truths and hoaxes. For experiments in truth, the result is a discovery. For demonstrations by a hoax, the result is an illusion. The people that study the paranormal understand the difference. Certainly The Archaeus Project group created its own truth-based understanding of the paranormal. The West Winds Mansion was and continues to be to this day a place to learn about and study the mysteries of life.

West Winds Mansion

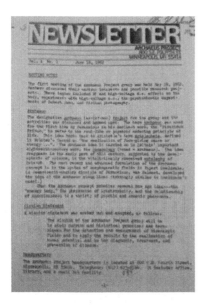 

Two of the Project's first newsletters.

Acknowledgements

As most projects there are a number of people and organizations that have helped create this story. I list them in no apparent order.
Nancy Blake
Champaign Public Library
The University of Illinois Main Library
The Bakken
Jade Beaty
Peggy Steffes
Ruth Mills
Karen Leivian

These e-books and paperbacks by T H. Harbinger can be found through Amazon books.

The Electricity of Life

A historical novel based on the true story of Otto and Viola Schmitt and their role in winning World War II and establishing a medical device industry

13 Stories From Living in Germany (English Version)

Whereas the most famous diary about Germany was from a young girl hiding from Germans these stories are all about trying to find them

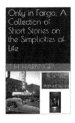

Only in Fargo: A Collection of Short Stories on the Simplicities of Life

Short stories on the daily life in the fictional version of Fargo. Everyday the simplicities and complexities of life create experiences that lead to the telling of stories, and then the writing of them.

Flying High With Consequences

A historical novel based on the true story of Professor John Karl and his research leading to the development of frac sand mines in Wisconsin.

Coming Soon

Gyro Landing

A historical novel based on the true story of the Ring Laser Gyro development, the sensor that aided in the Hudson River landing

Visit the authors website:
https://www.amazon.com/T-H. Harbinger/e/B00JEVD256

Proof

Made in the USA
Columbia, SC
04 August 2018